Pre-Industrial Economic Growth

Pre-Industrial Economic Growth

Social Organization and Technological Progress in Europe

Karl Gunnar Persson

Basil Blackwell

Copyright © Karl Gunnar Persson 1988

First published 1988

Basil Blackwell Ltd
108 Cowley Road, Oxford, OX4 1JF, UK

Basil Blackwell Inc.
432 Park Avenue South, Suite 1503
New York, NY 10016, USA

All rights reserved. Except for the quotation of short passages for the purposes of criticism and review, no part of this publication may be reproduced, stored in a retrieval system, or transmitted, in any form or by any means, electronic, mechanical, photocopying, recording or otherwise, without the prior permission of the publisher.

Except in the United States of America, this book is sold subject to the condition that it shall not, by way of trade or otherwise, be lent, re-sold, hired out, or otherwise circulated without the publisher's prior consent in any form of binding or cover other than that in which it is published and without a similar condition including this condition being imposed on the subsequent purchaser.

British Library Cataloguing in Publication Data

Persson, Karl Gunnar
 Pre-industrial economic growth: social
 organization and technological progress in Europe.
 1. Europe. Economic conditions, 1000–1800
 I. Title
 330.94′0146

ISBN 0-631-14963-5

Library of Congress Cataloging-in-Publication Data

Persson, Karl Gunnar, 1943–
 Pre-industrial economic growth: social organization, and technological progress in Europe.

 Bibliography: p.
 Includes index.
 1. Europe—Economic conditions—To 1492.
2. Technological innovations—Economic aspects—Europe—History. 3. Europe—Social conditions—To 1492.
I. Title.
HC240.P39 1988   330.94   88-5053
ISBN 0-631-14963-5

Typeset in 11 on 13pt Monotype Times Maths
by Advanced Filmsetters (Glasgow) Ltd
Printed and bound in Great Britain by Bookcraft Ltd, Bath, Avon

Contents

Preface		vii
1	**A Theory of Pre-Industrial Technological Progress**	**1**
	1.1 What Technological Progress is About	1
	1.2 The Stagnationist View Presented and Discussed	3
	1.3 A Theory of Endogenous Technological Change	7
	1.4 From Gathering to Cultivation	13
	1.5 The Early Metallurgy Sequence	21
	1.6 Mediaeval Agriculture and the Complementarity of Technological Sequences	24
	1.7 Conclusion	31
2	**Individualism and the Elementary Social Conditions for Technological Change**	**33**
	2.1 Introduction	33
	2.2 An Agnostic View on Social Institutions	36
	2.3 Feud and the Rituals of Exchange	42
	2.4 Scattered and Open Fields	45
	2.5 Guilds and Competition	50
	2.6 The Social Acceptance of Individualism	54
	2.7 The Problem of Institutional Change	57
3	**Growth and Stagnation in the European Mediaeval Economy**	**63**
	3.1 Introduction	63
	3.2 Common Ground and Interrelations	65
	3.3 What and When was Feudalism?	67
	3.4 Property Relations and Technology in Agrarian Transition	68
	3.5 Implications	70
	3.6 Urbanization and Population Growth in Europe	73
	3.7 Malthusianism Revisited	76

	3.8	Regional Diversity	78
	3.9	Incentives and Agrarian Transition	82
	3.10	Demographic Regimes and Economic Change	86
	3.11	Conclusion	88
		Appendix by Peter Skott	90
4	Measuring the Immeasurable: Labour Productivity in the European Mediaeval Economy		104
	4.1	Introduction	104
	4.2	Towards a New Approach	107
	4.3	Some Tentative Results on the Evolution of Labour Productivity	114
		Appendix	119
5	Why have Growth Rates been so Low until Recently?		124
	5.1	Introduction	124
	5.2	The Determinants of Technological Progress	124
	5.3	Labour as a Productive Force	128
	5.4	The Historical Significance of Piecemeal Technological Progress	129
	5.5	Continuity and Revolution	135
		Appendix	141
		References and Bibliography	143
		Index	153

Preface

This book is the product of an intellectual – and geographical – *Bildungsreise*. I have benefited from the company of the classical economists and Marx, mediaevalists and economic historians of many generations and modern economic theorists. I hope they will not find me too much of a free-rider. Although many of the problems addressed in the book have been tackled before there are new solutions as well as some new and intriguing questions asked and answered. So I think I have paid my dues.

My devotion to libraries has only increased during my working with the book. Anything less than devotion would in fact be impertinent considering the libraries I have chosen: the Royal Library in Copenhagen, the Bibliothèque Nationale and the British Library.

While in Paris I was provided with congenial accommodation on several occasions by the Centre Culturel Suèdois in the Marais and Mercurey at Aux Bons Crus. In London the Suntory-Toyota International Centre for Economics and Related Disciplines of the London School of Economics offered me the perfect intellectual and social milieu in which to do research (in 1985) and to write the book (in 1987). I owe the staff of these institutions a great deal.

When the book was in a formative stage I got decisive encouragement from my colleagues at the Institute of Economics, University of Copenhagen. I was privileged to have Jerry Cohen and Richard Smith, both of All Souls, Oxford, reading the draft of the book. Their enlightened comments improved the book considerably.

Bruce Campbell of Queen's University, Belfast, read chapters 3 and 4 and shared his intimate knowledge of the issues discussed in those chapters. Peter Skott, now at Aarhus University, wrote the appendix to chapter 3 and we have had many and long discussions on chapter 3 and the rest of the book. If this book still lacks rigour it is not Peter's fault. I thank them all for the concern they have shown for my work.

I have presented drafts of chapters to economic-history seminars at the universities of Copenhagen, Gothenburg, London (Birkbeck/LSE joint seminar), Lund, Oxford and Newcastle. On such occasions and in private conversations and correspondence I have got many valuable ideas and suggestions. Thanks to Hans Aage, Tony Atkinson, Meghnad Desai, Alain Derville, James Foreman-Peck, Christian Groth, Lars Herlitz, Lennart Jörberg, Barbro Nedstam, Agnete Raaschou-Nielsen, Claudio Rotelli, Gérard Sivéry, Claus Vastrup, Herman van der Wee and Tony Wrigley.

The Carlsberg Foundation and the Danish Social Science Research Council have generously supported me while working abroad.

To M.W.

1 A Theory of Pre-Industrial Technological Progress

1.1 What Technological Progress is About

Many contemporary economists, archaeologists and historians would not only question the possibility of building a theory about pre-industrial technological progress but also the rationale of such an endeavour. What is interesting and worth considering, so the argument goes, is rather the lack of technological progress up to the industrial revolution. That is not to say that technological progress is denied altogether but it is seen as exceptional rather than as an outcome of systematic forces in society.

One reason why this – what will be called – stagnationist view has been so widely accepted is that technological progress is believed to be necessarily associated with growth in per-capita product or income.[1] Although increases in income are far from absent in pre-industrial epochs it is of course true that per-capita income has been rising at a much faster rate from the industrial revolution onwards. That is, however, not a sufficient basis for a stagnationist argument since technological progress will not always cause an increase in per-capita product or income.

The general meaning that will be given to the concept of technological progress here is that a unit of a good or service is produced by a new technique using fewer resources than the previous one. A technique can be represented by a production function relating output to inputs. Technological change then represents a change in the parameters of the inputs in that production function. As a contrast, the concept of technical change is reserved for changes in methods of production involving only substitution of inputs but with unchanged parameters. In agrarian production, which attracts much attention in this book, we have experienced both technological and technical change throughout history. The former means that less land and/or labour is used in the production of a unit of, say, corn and the latter that there is a substitution of land for labour or vice versa in production.

The introduction of new goods and services is considered as a specific type of technological progress for reasons that need some elaboration. Take for example the use of iron in axes. It was once a new good and manufacturing of iron later undergoes technological progress in history. But how do we conceptualize its introduction? One way of treating it as an aspect of technological progress would be to consider the new good as an input into some composite production process, for example clearing of forests for agricultural production. Tools made of iron were more efficient than the ones made of stone, so in the production of a unit of corn fewer resources were needed, if we take both the resources spent on tool-making and the clearing of forests into consideration. Or take the case of the evolution of market services rendered by merchants and traders. The final effect of such services was that peasants could get involved in specialization that took advantage of regional comparative advantages in soil and climate, which increased agrarian productivity.

There are, no doubt, cases that are difficult to classify in the manner proposed above. Sometimes one may prefer to call a novelty a quality change that simply ensures more welfare. Similarly, when a product is new in the sense that one cannot identify a production process of a known good that involves the new good as an input or substitute, one must appeal to the contribution made of the new good to the ultimate good of welfare.

What are, then, the wider implications of technological progress? Often, but not necessarily, a growth in per-capita income. In general technological progress relaxes the constraints of an economy. If it is wrong to equate technological progress wth growth in per-capita product it is more appropriate to speak about it as permitting an increase in welfare or, more vaguely perhaps, as an increased ability to cope with natural and economic constraints.

Let us make the distinction between income (output) and welfare clear. Consider an economy that experiences technological progress so that a given amount of food is now produced by less labour time and land than before. Now if this society (i.e. members of it) chooses to go on producing the same amount of goods as before, it can increase the amount of leisure time per member of society. This is no less technological progress than if society chooses to work as many hours as before, permitting an increased production and consumption of goods.

Welfare can obviously increase by increased leisure as well as by increased consumption. But what if technological change is predominantly of a land-saving variety, which it was during much of the pre-industrial era? It is nonetheless formally sound and intuitively reasonable to view it as technological progress since land productivity is greater even if neither labour productivity nor leisure increases. Some of the constraints put on society by the scarcity of land are relieved. Technological progress with a land-saving bias makes it possible for a society to permit a larger growth in population than before since the scarcity of land is mitigated by better land use. It will permit societies to do away – at least to a certain extent – with the harsh methods of population control (such as infanticide) previously used. This is a far from negligible increase in welfare even though it is difficult to compare it with increases in income or leisure.

What we know for sure, however, is that there has been a continuous and remarkable saving of land, especially during the last 10,000 years or so, amounting to something like 500 times less land per capita for a subsistence consumption of a mediaeval peasant household compared to a hunter household in 10,000 BC.

1.2 The Stagnationist View Presented and Discussed

Although fairly widespread, the stagnationist view has no explicit theoretical foundation. Surprisingly, it is often based on *ad hoc* arguments. The archaeologist Colin Renfrew, whose innovative work has contributed to our understanding of technological progress in Europe, asserts nonetheless:

> Most preindustrial societies are in many ways conservative. They function successfully by carrying out traditional procedures whose effectiveness has been tried and proved over many generations. In order to survive, the society must to some extent function as a system that resists change, and all innovations, even potentially useful ones, tend to be viewed with suspicion. (1976, p. 205)

From another angle the economist John Hicks (1969, p. 13) argues that many preindustrial economies such as the neolithic village and the mediaeval village were customary societies based on rigid traditions of how to organize society and how to allot tasks to various groups of people. A similar argument is developed in a recent study of the novelty of the post-mediaeval accomplishment by the renowned economic historian Nathan Rosenberg (Rosenberg and Birdzell, 1986, ch. 2).

The dominant interpretation of European economic history from, say, 1000 AD to the industrial revolution associated with scholars such as H. J. Habakkuk (1958), E. Le Roy Ladurie (1966) and M. M. Postan (1966) is implicitly based on the idea of non-existent or very weak technological progress. Pre-industrial agrarian economies were therefore framed in a sort of Ricardo–Malthus trap. Periods of affluence stimulated population growth that eventually brought this type of economy to a standstill. Increasing population and the limited supply of land entailed rising rents for peasants, and agriculture was simultaneously experiencing declining marginal productivity. In chapter 3 this view is critically assessed.

The idea that pre-industrial economies are characterized by a stationary equilibrium is often associated with the view that they cannot experience substantial progress without some sort of exogenous shock. Various proposals as to what may constitute that exogenous shock are put forward, such as the emergence of new institutions, new mentalities, climatic changes or population-pressure-induced improvements in technology. Arguments of this kind tend to give undue importance to factors that are introduced in an *ad hoc* manner and therefore provoke more questions than are in fact answered. Why, for example, do societies and their members resist change?

There are no hints as to what the behavioural characteristics must be that are consistent with such claims. Neither an assumption of human beings as somewhat rational creatures nor an assumption of satisficing behaviour will, except under special circumstances, support the idea of inherent conservatism. Satisficing, as opposed to maximizing, behaviour only indicates that a certain target rather than a maximum is sought for. And such a target can of course provide an impetus to technological development. If insurmountable risks are associated with innovation then, of course, it would be rational to persist in using old methods. But if that is the explanation for the alleged conservatism, one must also argue convincingly that innovations associated with technological progress do necessarily involve great risks or uncertainty. While this may be true for some types of technological progress, I will argue below that this is not generally true.

Another argument sometimes put forward is that the obsession with material life and improvements, and increases in production of goods are cultural traits typical of the modern era. Pre-industrial cultures are, it is suggested, more inclined towards non-material aspects of life such as art, play, religion and ceremonies. Even if this is true – which is doubtful – it would not make these societies less inclined towards improvements in material production since thereby more time and other resources could be spent in the preferred non-production activities.

Although there are no convincing arguments for the existence of a conservative mentality as an inherent human trait, the stagnationist argument might still be plausible if the barriers to growth could be shown to be at work on the societal level. One possible candidate for such an explanation would be the proposition that innovation needs resources and that pre-industrial societies were too poor to devote sufficient resources to technological development. That point can be supplemented by the suggestion that there were no incentives for spending resources in technological development because the fruits of such endeavours were not channelled back to the innovators. In this view the lack of purposeful search for new technologies is not blamed on mental attitudes *per se* but on constraints in the form of material resources or institutional disincentives.

Not surprisingly, it is along these lines that modern versions of the stagnationist thesis are elaborated. Jon Elster, when criticizing the opposing view – i.e. the thesis of the development of the productive forces as defended by G. A. Cohen in *Karl Marx's Theory of History: A Defence* (1978) – touches this theme. Historical materialism, in its emphasis on technological progress as a typical feature of all human civilization, neglects, according to Elster,

> ...the institutional and psychological reasons that in many stretches of history have prevented technical progress, such as lack of investment objects (in the absence of a patent system) or lack of investment motivation (as in the non-diffusion of the water-mill in ancient Rome). Historical materialism, in Cohen's view, asserts an unbroken technical progress throughout history, even if proceeding at an unequal rate. An alternative view would be to say that the productive forces were largely static up to recent times, when the advent of capitalism made possible a revolutionary break-through.[2] (1980, p. 124)

This view is inspired by the so-called property-rights approach which stresses that the absence of appropriate institutions often hampers economic development if individual effort is inadequately rewarded. If, to take a highly relevant example, theoretical knowledge of a new and useful kind was freely available, i.e. if the originator of that new knowledge lacked property rights in it, then the originator created an externality, which is to say an uncompensated contribution to the welfare of others. Assuming that the originator was not primarily motivated by fame or intellectual curiosity, but by economic motives, then she/he would not be sufficiently rewarded to undertake the necessary work. The evolution of patent rights can be seen as a way of safeguarding necessary incentives for innovative thinking and experimentation, it is suggested.

D. North and R. P. Thomas have provided a more rigorous statement of this approach. In a way they spelled out clearly what many economic historians have long believed but not argued systematically before. Their main point is succinctly summarized in the opening paragraphs of their *Rise of the Western World* (1973).

> Efficient economic organization is the key to growth; the development of an efficient economic organization in Western Europe accounts for the rise of the West. Efficient organization entails the establishment of institutional arrangements and property rights that create an incentive to channel individual economic effort into activities that bring the private rate of return close to the social rate of return.

The general point that the growth of productivity and income in Europe was sparked off by institutional changes that liberated individual effort through an *internalization* of externalities by means of a (re)definition of property rights is challenged in this book. But that discussion is postponed to the next chapter.

On a more restricted issue I will, however, accept the ideas suggested by Elster as well as North and Thomas. It is true that pre-industrial economies, by and large, lacked institutions that rewarded purposeful search for technological improvements and for that and other reasons did not devote significant resources – to use the modern terminology – to research and development. Nevertheless I will not draw the conclusion that these institutional shortcomings prevented systematic forces creating continuous, though slow, technological progress. My argument is *not* primarily the idea that man in his natural curiosity is only slightly motivated by rewards and therefore indifferent to the institutional setting, although I think that this argument is worth some consideration. In the next section I shall, however, develop a more fundamental theme.

1.3 A Theory of Endogenous Technological Change

It is suggested here that there are important sources of technological progress that are endogenous in all production. Being a byproduct it will therefore have neither separable costs nor incentive problems of the type associated with technological progress seen as an outcome of purposeful search for new methods and deliberate experimentation. As a consequence, although intentional search for technological progress is weak, there will be systematic forces operating in favour of technological progress, and plausible positive feedback in the economic system may also generate self-sustaining growth. There are five basic elements in this process of technological progress.

1 Most productive operations are susceptible to certain random events or disturbances. During a long period with repeated productive operations, somewhat changed methods that originally occurred by chance might be chosen because they are considered superior by producers. An important characteristic of this type of piecemeal technological change is that it is costless and need not therefore involve the

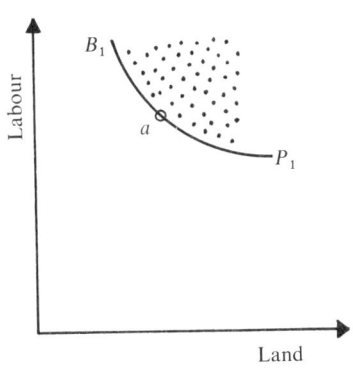

Figure 1.1

creation of modern institutions such as the patent system. Technological change of this type will be positively related to the number of productive operations made, which depends on the number of members in the economy and the level of per-capita income.

The argument can be discussed in detail if we consider an economy in which members use labour and land. In figure 1.1 a cluster of input combinations of land and labour for a unit of output is shown, all combinations represented by dots. This cluster of dots around an initial production method can be interpreted as the outcome of random variations in the original method. Assuming that people are somewhat rational, and possessed of both perceptive capacity and memory, we would expect them to single out those combinations that minimize the resources used for the production of a unit of output. In that process a best-practice curve $B_1 P_1$ might evolve, although members of a given society would not necessarily know its full shape, just certain points on it.

What is suggested is that producers in the course of history single out a production method, i.e. a specific combination of inputs of resources, on that best-practice curve. Any such recurrent and favoured input combination will be referred to as a standard method, for example point a on $B_1 P_1$. However, this standard method is – like the initial one – subject to small random events that are represented by a cluster of dots around the standard method a in figure 1.2. In due course, as experience is gathered and remembered, a new best-practice curve, $B_2 P_2$, will emerge and producers will find a new standard method b on that curve.

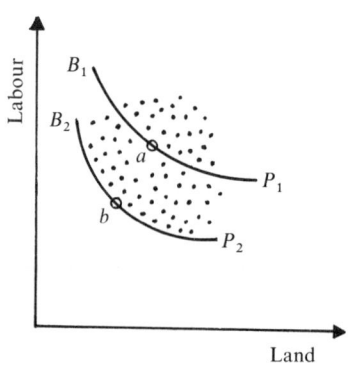

Figure 1.2

The inward move from points a to b in figure 1.2 is clearly a case of technological progress since resources are saved producing the unit of output with standard method b compared to a on B_1P_1.

If a best-practice curve, or a segment of it, is known we can also consider technical change as defined above. A movement along a best-practice curve, as illustrated in the figures, involves substitution of one input for another in the production of a unit of output. We consider such changes below.

2 There are in most productive operations what we can call *economies of practice*, which are the compound effect of learning by making a product (Arrow, 1962) and learning by using the product (Rosenberg, 1982, ch. 6). When a specific task is done repeatedly over a long period a producer will learn to master the job perfectly and to improve the product from experience in its use. This is also costless in the sense that it occurs as a byproduct of production.

To a large extent this is – as Marshall puts it – a process in which 'practice makes perfect' and therefore an inalienable property of the individual producer. This will in fact create incentives for producers to get involved in specialization since the learning of skills will be associated with some concurrent personal advantages. Learning by using will manifest itself as changes in implements but the necessary knowledge has been gained costlessly. Economies of practice can also be represented as an inward shift of the preferred standard method as illustrated in figure 1.2 from a point at B_1P_1 to a point at B_2P_2.

Although much of the perfection is tied to the personal experience of the producer – essentially dependent on the time that the producer has devoted to a specific task – part of the accumulated knowledge will be transferable. This is particularly true of the changes in the design of implements and tools caused by learning by using. The diffusion of knowledge that is not embodied in implements can be made more or less costless through demonstration in production. As long as this is done from generation to generation there are no disincentive effects especially if that transfer of knowledge takes place within the kinship or family, which typically is the case in agriculture. An incentive problem may arise if there is *intra-generational* transfer since an apprentice can become a potential competitor. The peculiar institutional arrangements that surround apprenticeships, including restrictions on entry into professions, may be interpreted as an attempt to reconcile the problem of disincentives in teaching and advantages in learning skills.

Institutions that facilitate transfer of knowledge will create a technological heritage for a society.

3   Trial and error can be considered a conscious extension of knowledge about production techniques, and costs are negligible if trials are small variations in established methods of production. The effects of trial and error are very similar to that of selection from random disturbances in production methods as described in point 1, except that there is an intentional element in trial and error. The costs of trial and error should not, however, be exaggerated. It might be part of leisure and play in pre-industrial economies as it is to some extent in modern society. Even though some costs are involved in trials there was no immediate need for a patent system in pre-industrial societies. A patent system is supposed to channel rewards by giving temporary rights of a new implement to the innovator who could charge others for using the new knowledge. In most pre-industrial economies, diffusion of new inventions was so slow that innovators got their temporary rights by remaining the sole user of new knowledge for long periods.

4   Division of labour and regional specialization are two interrelated sources of technological progress that are both enhanced by population growth and the growth of markets. Division of labour occurs when an integrated productive operation is separated into specific tasks each performed by a specialized producer. There are at least three distinct reasons why division of labour increases productivity. One has already been mentioned when discussing the fact that economies of practice are enhanced by the number of times a specific task is repeated by a single producer. Division of labour will make the typical producer concentrate on fewer tasks thereby increasing the productivity in the execution of the remaining ones performed more often and with increasing perfection.

Division of labour is usually associated with the learning of skills and the introduction of new tools or equipment, both exhibiting indivisibility characteristics. Indivisibilities in learning (equipment) imply that production methods have minimum requirements of training (capital) to make efficient production possible. If much training is needed or if the equipment is costly a production method may not be practised unless a considerable demand ('the extent of the market', in Adam Smith's words) exists.

The problem just raised can be illustrated by figure 1.3. Imagine the production of a good, Q, with labour, L, as the only input, and two

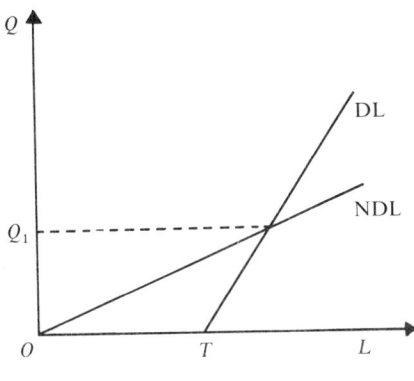

Figure 1.3

types of technologies called NDL (no division of labour) and DL (division of labour). The DL technology exhibits indivisibilities in learning measured by the labour time necessary for the training of labour (OT in figure 1.3). As long as demand is smaller than Q_1 the NDL technology is more efficient than the DL, i.e. the labour productivity Q/L is higher for the DL technology than for the NDL technology. Demand above Q_1 will make the DL technology pay off the indivisibilities in training since labour input (both training and production) is now lower per unit of output compared to the DL technology.

While it can be safely assumed that the most efficient known NDL technology is always used, there may or may not be DL technologies available which are more efficient but not actually used because of insufficient aggregate demand. Insofar as there are superior technologies exhibiting indivisibilities, an increase in aggregate demand (the extent of the market) will, however, remove the barriers to their use, causing an immediate decrease in unit costs.

5   The role played by population growth now becomes evident. In economies where per-capita income grows slowly, the extent of the market will be dependent primarily on the population density. Increasing density increases the extent of the market as will development of transport technology, forging isolated economies into larger entities. Increasing population will augment aggregate demand and relieve the economy from some of the barriers posed by indivisibilities in learning and equipment. To the extent that continued population growth is possible we may experience a type of self-sustained growth.

One requirement is that technological change is land-saving, though not necessarily exclusively so. This is plausible because an increased division of labour also implies the emergence of markets, which will stimulate regional specialization. By exploiting regional comparative advantages, land-use efficiency will be enhanced. A full account of the conditions for self-sustained growth is discussed in chapter 3.

Although there is an anti-Malthusian element in this argument it is different from the fashionable population-stress hypothesis, which is discussed below. In the approach favoured here, population growth stimulates technological change through its effects on the division of labour and regional specialization. Since economies of practice are related to the degree of division of labour, there will also be positive learning effects affecting productivity growth.

The argument pursued here identifies strong endogenous forces in the process of technological development. These forces generate what will be called *technological sequences*. Of major importance is that a technological sequence can be considered to have a deterministic trajectory. If we consider a specific activity, such as the methods that apply to the cultivation of land or metallurgy, the suggested determinism predicts a certain evolution of technologies over time. The basis for this assertion is the fact that given a specific standard method we will consider the evolution of the technology as resulting from a selection from small variations close to the original standard method. The essential characteristic for a preferred method is that it saves resources. With the plausible assumption that societies at comparable levels of development face similar constraints as far as resources are concerned, it is reasonable to suggest that their technological trajectories follow similar, but not necessarily identical, patterns. In the same vein, it can be argued that the process of division of labour follows a similar evolution across cultures because the possibilities of the fragmentation of the production process in subspecialities are entailed in the technical characteristics of production. These are important implications at odds with much of the traditional archaeology that tends to favour a diffusionist approach of major technological events.

The argument pursued so far boils down to the assertion that important technological sequences, such as the sophistication of metallurgy or the beginnings of agriculture, have emerged and developed in broadly similar ways in many different cultures independently. And, furthermore, if a culture imitates the technology of another culture at a certain stage it will be able to develop along much the same trajectory as the original innovator. We will return to this crucial implication when discussing comparative technological history later in this chapter. This will be done by discussing technological sequences, their origin and internal logic in different historical epochs and in varying areas of production. First the transition to agriculture, second the early metallurgic sequence and finally the interplay of different technological sequences in the transformation of mediaeval agriculture.

1.4 From Gathering to Cultivation

No-one denies the far-reaching consequences of the emergence of agrarian civilization as opposed to nomadic hunting and gathering societies. The very sedentary nature made possible by agriculture enhanced inter-generational cultural continuity and, consequently, the accumulation of a technological heritage, which is handed over through the evolution of technological sequences. Furthermore, the increased density of population, which also had agriculture as an indispensable prerequisite, generated the necessary conditions for division of labour and regional specialization.

The origins of agriculture are, however, hotly disputed. The view presented and defended here is of an *endogenous and independent evolution* variety as opposed to a strong tradition suggesting an *exogenous shock* and aften, but not necessarily, *cultural diffusion*. More precisely, it will be suggested that cultures practising gathering of plants gradually developed a knowledge that in due time led them to try to cultivate selected plants on land suitable for that purpose. Gathering and early agricultural trials can therefore be seen as part of a technological sequence in which purposeful cultivation replaces gathering once sufficient knowledge about the environment has been accumulated. Once agriculture has reached a level of a standard practice it would be wise to shift the focus somewhat. At that stage it makes sense, from an analytical point of view, to consider agriculture as a sort of meta-sequence consisting of a series of sequences each undergoing evolution such as techniques related to efficient land use, traction power and crop selection (cf. section 1.6 for elaboration of that point).

The 'gathering to cultivation' sequence certainly developed slowly and it is probable that there were intermediate forms between nomadic life and stable sedentary life with continuous cropping involving rotation, fallow and manuring to keep the soil in good condition. G. Clark and S. Piggot (1970, p. 161) suggest the following intermediate type of 'protoagriculture' although they do not use that terminology. They suggest that some early tell formations may be interpreted

> ... as perhaps the result of recurrent reoccupation of the sites in a cyclic rhythm that allowed for natural regeneration over perhaps a relatively large number of years. It seems scarcely credible that from the very beginning of animal and plant husbandry in the Near East a stable form of completely permanent settlement should have been achieved out of a mobile hunting-and-collecting economy virtually overnight.

In fact such semi-sedentary life might be the way societies learnt some basic advantages of agriculture but they certainly had a long way to travel before they became full-time agriculturalists. Likewise proto-domestication of sheep, pigs and cattle possibly got its impetus from the deliberate or accidental emergence of browse and selective clearance of land that attracted animals, out of which sophisticated ways of domestication later developed (Whittle, 1985, p. 19). Both wild cereals and pulses have been found close to early agrarian cultures, but it is probable that the knowledge that pulses can replace some of the fertility lost in growing cereals has been gained by chance, as possibly was the case with the advantages of manuring. Similarly, rotation and fallowing regimes have taken a long time to develop as have the morphological changes in cereals through selection in agrarian practices.

The very slowness of evolution makes it less likely that agriculture was a response to some immediate crisis, be it population pressure or climatic changes, as suggested in those explanations that rely on some sort of exogenous shock as the driving force. The idea that a culture gradually extends its knowledge about the environment, and learns how to master it and manipulate it has also been advanced by some archaeologists as regards the transition to agriculture – most clearly by R. J. Braidwood, whose position has evolved over a long time and partly in response to the dominant 'exogenous shock–cultural diffusion' paradigm. There is no need to complicate the story with exogenous causes but rather interpret the transition as the 'culmination of the ever increasing cultural differentiation and specialization of human communities'. The agricultural transformation simply occurred when the inhabitants of an area knew their environment so well that they could cultivate the plants and domesticate the animals that they previously collected and hunted (cf. Braidwood and Howe, 1960, pp. 2–8, and Braidwood and Braidwood, 1969).

The simplicity of this explanation has not, however, attracted the support it deserves. The fact that diffusionist explanations so often have been preferred – more often a decade ago than at present however – is, I believe, because there has not been an explicit acknowledgement of technological progress as endogenous in production. For that very reason some sort of exogenous shock is considered necessary to get allegedly conservative societies on the move.

Explanations that attribute an important role to an exogenous shock remain, however, in the forefront and therefore merit closer inspection. The eminent archaeologist V. G. Childe formulated one of the most influential with what can be called a climatico-ecological view. Agriculture appeared for the first time in the latest post-glacial period and it is suggested that climatic and environmental forces acted, directly and indirectly, as the prime movers. In west Asia, desiccation enforced a juxtaposition of men, animals and plants suitable for cultivation in oases and river valleys that ultimately promoted the adoption of agriculture. The chronological record showing that west Asian agrarian sites predate most European sites originally stimulated a diffusionist explanation of the emergence of agriculture on a wide scale. The most simplistic is the Ammerman and Cavalli-Sforza wave-of-advance model measuring a westward advance of agriculture at a speed of almost 20 km per generation starting some 8000 years ago. The chronology of agrarian transitions is in itself not sufficient proof of diffusion and the actual mechanism of diffusion – i.e. immigration, cultural diffusion or imitation – is seldom discussed explicitly. The view that agriculture was introduced through immigration of agriculturalists pushing back hunter-gatherers has recently been disputed (Phillips, 1981, pp. 152–5).

The climatico-ecological view has been challenged primarily because agriculture has appeared under a wide variety of climatico-ecological conditions throughout the world and some critics even raise doubts concerning the validity of Childe's hypothesis as applied to west Asia (Braidwood and Howe, 1960).

Interpreted more generously, the importance of the climatico-ecological changes can, however, be integrated in the endogenous technological progress approach. Environmental changes do not of course impose a certain course of action directly on man but a new set of ecological conditions provides man with new experience. It would certainly help societies make the transition to agriculture if an ecological metamorphosis made plants suitable for cultivation appear in great numbers in the vicinity of populated sites.

Explanations that refer to the active role played by population pressures have replaced the climatico-ecological approach in recent discussion, largely as a result of the publication of E. Boserup's book *The Conditions of Agricultural Growth* (1965). The theme elaborated in that book is that man substituted labour for land due to population pressure and increasing scarcity of land. Agriculture occurs – in this view – when traditional hunting techniques become as labour-intensive as agriculture because more people compete for the shrinking food available for hunting and gathering (Boserup, 1981). Agrarian techniques are known but not used until the labour input for a unit of food is less than in the traditional hunting activities. In her first book, E. Boserup was concerned with changes in agrarian techniques towards an increasing efficiency in land use, and the adaptation of her argument to the problem of the agrarian transition was suggested by M. N. Cohen (1977), of which she has approved in a recent book, however (1981, ch. 4). It could be useful to quote Cohen at some length:

> Population pressure must be redefined. *It is here defined as nothing more than an imbalance between a population, its choice of foods, and its work standards, which forces the population either to change its eating habits or to work harder* (*or which, if no adjustment is made, can lead to the exhaustion of certain resources...*). This altered notion of population pressure, incidentally, answers one criticism which is often levelled at the concept of population (or stress in general) as an explanation of the origins of agriculture. It has been argued ... that agriculture could not arise out of food shortage, because people under stress do not innovate; they lack leisure time for experimentation with new techniques and cannot risk their scanty supplies on the uncertain promise of distant future rewards. It has already been pointed out, however, that experimentation probably has nothing to do with it. The techniques are already widely known; all that remains is to implement them. (1977, pp. 50–1, author's italics)

The basic problem with the population-stress hypothesis is that it fails to convey a plausible mechanism that transforms population pressure to agricultural change. There are, no doubt, attempts to provide such a causal link but they rely on questionable assumptions. First, it provides pre-industrial man with a rather restricted behavioural strategy; second, it does not seriously consider the view that agricultural change is a case of technological change; and finally, it tends to play down the rather sophisticated knowledge needed for the transition. The final issue has already been dealt with above so we can turn to the first problem.

It is supposed that labour effort is dependent on an exogenously given population growth and a land constraint. The principle governing labour effort can be illustrated by going back to figure 1.2 and interpreting the inputs of labour and land as expressed in per-capita terms. Assume that population and land availability permit production at a. This is a point at which the least possible effort is put into production given the exogenous constraints: technology, population and availability of land. Only continuing (and uncontrollable) population growth can – according to the population-pressure hypothesis – make society produce at a point to the left of a on $B_1 P_1$. Most societies seem to have practised some sort of population control, however. Boserup and others have countered this critique by arguing that population-control regimes may for some reason break down (temporarily) or be non-existent in some societies, and these are the ones that adopt agriculture (Binford, 1968; Boserup, 1981, p. 39).

This is but conjecture, of course. The population-stress hypothesis is weakened further when we consider the nature of the labour-supply assumption. It is implicitly assumed that the actual labour effort offered is larger or equal to the preferred level of effort. This is not necessarily true in pre-agrarian societies because one of the most impeding conditions of such societies is the inability to permit or choose effort at a desired level. A hunter–gatherer economy is basically a parasitic culture. What is believed to be a preferred behavioural strategy is imposed by ecological conditions. Any increase in labour effort when the society is in ecological equilibrium would tend to disturb the reproductive patterns of the species upon which the human populations depend.[3] Deliberate cultivation and domestication of animals made it possible for a population to vary its labour effort to a desired level taking other goals such as the size of the population and the level of output into consideration. This is exactly why it must be considered an increase in welfare: one important constraint – the exogenously determined labour effort – was eased with the transition to agriculture.

The argument pursued has important implications for the interpretation of the agricultural transition. It is perfectly sound to argue that societies embarked upon agricultural attempts because they wanted to feed a larger population, although they may or may not have been forced to work more to gain these advantages. This is not a population-pressure-induced transition to agriculture but an increase in population because of the increased production made possible by the adoption of agricultural methods. In archaeological research there is, unfortunately, no way one can tell whether an increase in population was an effect of the introduction of agriculture or caused that transition (see Barker, 1985, pp. 259–60).

Not only does the population-stress hypothesis use cumbersome behavioural assumptions, but it also tends to view the transition to agriculture as a matter of technical change rather than a genuine technological change. The former restricts the choice facing producers to a matter of substitution: typically labour for land, i.e. a leftward movement along a production function such as $B_1 P_1$. With the rigidity principle governing labour effort we are not surprised to learn that producers will not embark upon labour-augmenting agriculture just to save land unless they are forced to do so by population growth. If we admit technological change, i.e. an inward shift of the production function, the problem will not be one of substitution but of saving both labour and land per unit of output. If the agrarian transition involves technological change as opposed to mere substitution of resources there is no need for population pressure to make people accept it. The role of population pressure can be reduced to the equally important one of stimulating a bias in the selection of new knowledge so that land-saving technological change is fostered.

It seems beyond doubt that the transition to agriculture caused substantial increases in labour productivity. Part of it may have its origin in more hours of labour being spent in production. As argued above one truly innovative characteristic with non-parasitic cultures is that they can vary their labour effort to preferred levels which may affect man-year productivity positively, and when we speak about labour productivity it is precisely man-year productivity that is considered. Is it appropriate to consider growth in labour productivity as a type of technological progress if it depends only on an increase in hours worked? I believe so, but only if the additional labour can be expanded because some previous ecological or natural constraint has been erased. It can thus be interpreted as an increased efficiency in the use of a resource – labour in this particular case. The view that the low level of effort observed in some surviving hunter-gatherer societies is a sign of affluence is obviously not accepted here. It is more likely an exogenous constraint – see Sahlins (1974) for the opposing argument. It is first with agriculture that advanced culture appeared: the evolution of towns, intellectual and professional specialization, trade and the emergence of larger territorial units. Given the plausible assumption that the masses did not experience a decreasing standard of living because of the transition to agriculture, such a cultural evolution and differentiation is impossible unless labour productivity in food production has increased.

D. North (1981, pp. 72–89) has put much of the reasoning by archaeologists and anthropologists reviewed above on an explicit theoretical basis. His argument is based on the premise that agriculture was a known technique which could be adopted simply by substitution of inputs. The whole story of agricultural transition can then be told as a type of technical change: relative scarcities of different inputs changed as a response to population growth. Thereby, the marginal productivity of labour in hunting and gathering diminished because of the overexploitation of the game, making agriculture compare favourably. So far North does not add anything to the conventional view except a certain amount of rigour and a new terminology. He also criticizes, however, the population-pressure approach for not having an explanation for the alleged uncontrolled population growth. The hypothesis he offers has some interesting properties.

Presupposing that several groups compete for the same game in the same territory, two consequences can be deduced which both emerge from institutional failures. First, there are only weak incentives for a group to restrain population growth, because if it did that particular group would be outnumbered by competing groups. Second, there is an inherent tendency to excessive hunting if several groups compete for the same game since it is not in the interests of a single group to limit its hunting if others do not do likewise.[4] As long as the game is common property, such behaviour is difficult to monitor. This institutional failure entails the exhaustion of resources and ultimately makes agriculture compare favourably with hunting and provides the rationale for the transition.

The weak link in this hypothesis is the assumption that hunters and gatherers did not develop any sort of exclusive claims on territory. It would perhaps be more costly but equally advantageous for hunters as for agriculturalists to establish property rights or some sort of collective regulation of access to the resource, private or public. The costs of supervision of territory should not be exaggerated, however, because it is a byproduct of production: when hunting the area it is also supervised. Archaeological research cannot tell us whether primitive man did or did not in fact establish private or collective property rights. Surviving groups of hunters seem to have exclusive rights in territory, especially if there is competition for resources. The conjecture advanced here is that such collective property rights developed before the transition to agriculture. The basis for that conjecture is that agricultural transition is the outcome of a growth of knowledge process that presupposes a cultural continuity over a large number of generations. The archaeological record tells us about fairly stable protoagricultural and foraging societies. In the absence of group rights over territory such continuity would have been disturbed repeatedly because of over-exploitation of resources. This may be a reason why pre-agrarian cultures lasted so long. The very instability, for institutional and other reasons, did not permit the accumulation of sufficient knowledge until feasible institutions provided the pre-conditions for cultural continuity.

It was stressed above that the endogenous technological progress approach predicted independent (or multi-centre) evolution of broadly similar technological sequences. By and large, the diffusionist view according to which Old World agriculture spread from a centre in west Asia has fallen into disrepute in recent decades. The quarrel now is not whether there were independent agricultural transitions but how many there were. While it has always been clear that agriculture in the Old and New Worlds developed independently, the focus is now on the identification of independent centres of agrarian breakthrough in Europe and Asia. That there were several is clear; that there were many is probable. Concerning Europe, G. Barker (1985) interprets the available archaeological data as implying an independent evolution in the Mediterranean basin but stimulated by the exchange of emmer and bread wheat with west Asia. Similarly, he rejects the idea that agriculture in the Balkans and the middle Danube basin was introduced by Greek colonists. A sort of protoagrarian culture practising foraging on a large scale was identified before 5500 BC which suggests an independent agrarian origin (Barker, 1985, pp. 97–8). The continental lowlands in Europe are believed to have experienced an independent agrarian transition while its evolution in the British Isles '...can probably be understood just as much in terms of internal changes in the existing foraging economy as in terms of a system imposed or introduced by new people...' (Barker, 1985, p. 203).

Another recent survey of archaeological data for Europe by A. Whittle is more cautious, talking about the probable indigenous development of agriculture in the areas just mentioned (Whittle, 1985, ch. 8, and also pp. 54–5). Several independent agricultural transitions in the Mediterranean regions are admitted as plausible, however. Possible candidates are both Italy and Spain, although available data are not conclusive. Whittle confirms the view that the mono-centre conception of the origins of agriculture must be replaced by a multi-centre approach. A similar change of approach is visible in the analysis of the evolution of metallurgy, to which we shall now turn.

1.5 The Early Metallurgy Sequence

Metallurgy provides a very instructive picture of the pattern of a technological sequence as being the result of selection from chance variation and economies of practice that gradually generate a technological heritage, part of which can be transferred from one generation to the next. There is also evidence of deliberate experimentation by the craftsmen. Before presenting a short outline of that process let us look briefly at the geographical location of major metallurgic innovations. As in the case of agriculture there has been a strong diffusionist tradition viewing European metallurgy as imported from west Asia, although the independent evolution of New World and Chinese metallurgy of course has been admitted. There were probably several independent centres of metallurgy in east Asia.

C. Renfrew, who in a number of works has attacked the diffusionist view in the light of new chronologies that are based on radiocarbon datings, has also vigorously defended the independence of European metallurgy (1969, 1976). Recent research confirms that there probably were several independent centres of early development in Europe (Phillips, 1981, pp. 182–3).

Innovative centres in metallurgy were located in areas with old permanent settlements of mixed farming, i.e. areas that exhibited cultural continuity and having, presumably, an accumulated technological heritage in several lines of production. There are of course strong theoretical arguments for the multi-centre view of technological innovations apart from the empirical ones. Metallurgy can be seen as an outgrowth and fusion of two technological sequences: the pyrotechnical knowledge obtained from ceramic production, and a sequence related to the manufacturing of tools out of stone. Stone technology, the predecessor of metallurgy, was not in itself without considerable progress but increases in efficiency in stone tools were bounded by the characteristics of the raw material. The traditional way of manufacturing was restricted to slate, but when grinding of stone was learnt, the variety of useful raw materials was greatly extended. The efficiency of axes and adzes was enhanced by grinding which promoted forest clearance and slash-and-burn agriculture. But stone had obvious limitations as a raw material, most notably the difficulties in making tools with sharp edges. The edge angle was progressively reduced in history but with stone it seldom fell below 50°, most axes having edges somewhat less sharp. Early introduction of metals reduced the angle on an axe blade to 15–20° (Semenov, 1964, pp. 200–6).

The first metallurgic practices were entirely mechanical, primarily hammering and annealing of native copper and gold. Such experience can be seen as a continuation of skills developed in stone technology, i.e. selection of appropriate raw materials and mechanical fabrication of objects for use. Whenever mineralization was evident we would expect stone-age cultures that had gained sophisticated knowledge of their environment to try to use these materials. Hammering and annealing restrict the range of objects that can be manufactured, however.

It is with the use of heat that the development of metallurgy for productive purposes gains momentum. A basic pyrotechnical knowledge was present before the advent of metallurgy through ceramic manufacturing. It has been suggested that melting had in fact first been discovered by chance with native copper in kilns. Another possibility – not excluding the one just mentioned – is that the knowledge that flintwork was facilitated when stone was heated inspired primitive forging of native copper. In this process it is plausible that it was learnt that copper melted at high temperatures. Once melting was practised, it was a short step to using alloys of tin and copper or arsenic and copper to make bronze, with its superior properties for implements in war, pleasure and production. Melting of alloys also invited man to experiment with casting which greatly increased the range of objects that could be manufactured, for example axes and adzes (Charles, 1980, p. 168).

Further development of metallurgy increased the range of raw materials that could be used, profiting from a refined pyrotechnical experience. At an early date of the metallurgy sequence it becomes reasonable to speak about furnace technology and the ingots produced were fairly standardized. With furnaces, complex ores could be used provided the reduction process involved a suitable flux, iron oxides being preferred. It is sometimes suggested that iron was discovered as a useful metal when it was used as a flux and emerged as a byproduct when producing copper through a reduction process. Be that as it may, the knowledge of reduction of ores was necessary for the subsequent development of iron technology. Schematically speaking, iron technology was based first on a gradual improvement of pyrotechnical knowledge so that increasingly high temperatures could be mastered, and second on refined control of the carbon content of the product. Reduced iron was taken from the furnace as a bloom that with additional hammering yielded iron that could be smithed. With appropriate heating, requiring bellows, the iron could absorb so much carbon that cast iron was produced (Tylecote, 1980, p. 209). Wrought iron dominated practical uses of iron in Europe up to modern times while the Chinese were very quick to adopt cast iron. Not until mediaeval times, when European metallurgy developed considerably in terms of production and techniques, did the technological gap begin to narrow. Although there were regional variations and the rate of progress differed – the New World did not independently discover iron, while sub-Saharan Africa started its impressive metallurgic sequence with iron – the general sequence can be fitted into a growth of knowledge framework: a 'sequence of increasing competence' as Renfrew prefers to call it.

The reason why the New World did not reach the iron stage might have to do with its comparatively short independent experience with metallurgy, approximately two millenia. In the Old World some four to five millenia elapsed from early trials in metallurgy until iron was manufactured systematically in the last millenia BC.

The simplicity of the conceptual framework used here is necessarily associated with the high level of abstraction. Regional studies and closer chronological research will certainly reveal interesting peculiarities which are caused by factors outside the simple logic of a technological sequence. This will not, however, fundamentally alter the fact that the pattern is one of slow progress. Why technological progress was so slow is the subject of the last chapter.

1.6 Mediaeval Agriculture and the Complementarity of Technological Sequences

By and large mediaeval agriculture has been judged by its weaknesses rather than by its accomplishments. Its most obvious weakness was the vulnerability to exogenous shocks such as war and social disorder, unfavourable weather and climatic changes. A succession of years of wet and cold weather had grave consequences because stocks were depleted and international trade was still not sufficiently developed. Harvests were often reduced to half the normal level in years of bad weather.

Also in this respect, however, mediaeval agriculture made some progress. That great transformation of land use known as the transition from a two-course rotation to a three-course rotation system not only implied a declining proportion of land in fallow. It also stimulated the combination of autumn- and spring-sown crops. This diminished the risks of harvest failure created by an unpredictable climate.

But there were other important technological transformations both in terms of diffusion of known techniques and evolution within several technological sequences. The period that witnessed most of these transformations was the era stretching from the eleventh century to the Renaissance.[5]

Some of the technological improvements are just refinements of existing tools and might at first sight seem so small that they are not worth consideration. But as has been pointed out by Charles Parain (1979) – a pioneering researcher on mediaeval technology – many such small modifications add up to remarkable improvements in the implements used in production.

Scythes and sickles, to take just two examples, went through a process of small changes in design that made them more efficient. In ploughing, to take another example, the labour requirement was also reduced so that one person managed both the draught animals and the plough. The process of piecemeal change is well described by G. E. Fussel (1966) discussing ploughs:

> Slow and unrecorded developments must have taken place in the 500 years after the Domesday book was written. No doubt these were mainly slight changes in the details of the generally accepted design of ploughs. Many, if not most, of them have been made at the request of individual farmers and were dictated by experience.

Progress in late mediaeval agriculture has been viewed as the outcome of several seemingly independent sequences, but in fact there was interdependence between sequences. The development within a sequence is looked upon largely as a spontaneous process, i.e. a growth of knowledge gained in production rather than by deliberate search. One implication is that complementarity of development in different sequences is far from perfectly synchronized. We should expect bottlenecks and delayed complementarity. A lack of synchronization created by the relative retardation within a specific sequence can, however, create a bias in the selection of new knowledge from experience.

The problem of complementarity can be illuminated by the connection between the *traction-power sequence*, which relates to the exploitation of power from draught animals, and the techniques bearing on the efficient use of land – the *land-use sequence*. More efficient use of draught animals was complementary to the introduction of the deep-cutting heavy plough with mould-board which opened up new types of fertile soils to cultivation and accompanied the transition from a two-course to a three-course rotation, or more generally the gradual suppression of the fallow. As a consequence, the centre of the European economy moved from the Mediterranean to Central and Western Europe once implements and methods were adapted to the specific demands of climate and soil.

The more permanent use of land as well as the wider variety of soil types that could be sown was complementary to a *crop improvement and specialization sequence*. In this sequence, specific crops were tried and improved (through selection) for the characteristics and peculiarities of soil, climate and demand. With the interdependence of different sequences in mind, a discussion about each of them to demonstrate the logic of their own evolution now follows. It should be pointed out, however, that the following short survey is neither detailed nor complete. The most important omission is perhaps an account of the accompanying evolution within metallurgy. It can be suspected from what has already been noted in passing about ploughs, scythes and sickles that there was an increase in the production and use of iron. With the application of water-powered hammers and bellows there were economies of scale leading to falling prices which stimulated the use of iron. Apart from the uses already mentioned the following can be added: the nailed horseshoe, tires, the iron-reinforced spade, the clock and many of the increasingly specialized tools that the urban craftsmen developed. The general trend, described in the section on the metallurgy sequence, towards mastering the heat in furnaces continued, notably with the new, more efficient bellows. The shape of scythes and sickles tended to become more appropriate, but also more demanding in terms of the quality of the iron and the skills required by smiths.

The Land-use Sequence

Throughout history man has struggled to economize on land. As has been noted already the agrarian transformation was in itself an important step in that direction. But agrarian techniques differ widely in their requirement of land. Early agriculture is characterized by extensive use of land of the slash-and-burn variety with fallows ranging from ten to 30 years. In the early mediaeval era, a two-course rotation system had been established, and the ensuing centuries witnessed the transition to a three-course rotation and, in some parts of Europe, the almost complete suppression of the fallow.

In the late mediaeval era, soil use was thus intensified and great ingenuity was needed in restoring and maintaining the fertility of the soil. Throughout the period there was improvement in many practices including ploughing, fertilization, crop rotation, weeding and regulation of the humidity of the soil by means of irrigation or digging (with the iron-reinforced spade) of ditches. The intensive use of land not only increased the yield per land-year (i.e. the yield per unit of land per year) but, more surprisingly, yield per seedcorn as well. From yield per seedcorn as low as 2.5:1 in early mediaeval agriculture, yields tripled or quadrupled well into the late mediaeval period. We would expect that an intensified use of land would lead to the impoverishment of the soil. Although such tendencies are recorded, they are far from general. Repeated workings of the soil with the mould-board plough accompanied other improvements in the conservation of land quality. The problem solved by that type of plough was specifically posed by the heavy soils of temperate Europe because there was a need to cut deep and turn the cohesive sods.

The task of cultivating the heavy soils of Europe did, indeed, start before the late mediaeval era but it culminated in that period. The plough with mould-board must be seen as a superior solution to a problem facing peasants at earlier stages of agrarian history. More specifically, the early ploughs forced the ploughman to turn the sod by his own power. With the new plough, the power needed for turning the sod was assigned to the draught animals (Jope, 1956, pp. 86-91). It is worth noting that peasants developed a fine knowledge of the varied requirements demanded by different soils. This knowledge included when, how and how often land should be ploughed and manured and to what uses specific soils could be put.

Although the advantages of manuring were well known before the mediaeval era, it was intensified and new methods were introduced or rediscovered such as marling and green-manuring. The increased urbanization supplied areas close to cities with night-soil that supplemented the manure from livestock. The increased tendency to stall livestock facilitated the productive use of manure, and in some areas convertible husbandry – normally ascribed to the agrarian transformation several centuries later – was practised whereby land was transformed temporarily to pasture for the livestock.

High labour input substituted the declining land available per capita and was one of the reasons why land could be used more intensively. In the most advanced areas, rotation schemes included pulses often sown in the fallow year that added much-needed nitrogen and produced fodder for the livestock, and consequently manure. Elaborate rotation schemes also impeded the exhaustion of the soil because the root systems of different plants do not penetrate the same layers of the soil. Furthermore the rotation of crops hampered the spread of parasitic weeds and insects since they often take a specific plant as host.

All these practices, pursued with differing skill and intensity across Europe, bear witness to a considerable growth of knowledge. Of course none of the beneficial effects on soil fertility were understood theoretically. It must be seen as a piecemeal accumulation of practical experience. The fact that Europe saw the parallel evolution of several centres of agrarian development, notably the Low Countries, northern

Italy, the Rhine valley and south-east England, can be taken as support for the hypothesis of independent and endogenous mechanisms of change. On the other hand, since those areas leading the agrarian transformation were also the most active participants in the flourishing international trade, a diffusionist approach gets its share of support as well.

Additional support for the view that there were several foci of independent agrarian progress comes from the interpretation of land intensification in agriculture as a spin-off effect of practices in gardening (van Houtte, 1980, p. 45). In the cultivation of fruits, vegetables and industrial plants, permanent land use was the rule with considerable input of labour and natural fertilizers. Such gardening cultures developed wherever cities grew sufficiently large to become a source of demand.

Over the centuries of agrarian transformation the compound effect in terms of increased intensity in the use of land was tremendous, but it is important to note that it did not consist of a few technological leaps. On the contrary, if we deconstruct the transformation we will recognize a long chain of small improvements which can be explained perfectly well as the outcome of endogenous change in a technological sequence.

The Traction-power Sequence

The slow but continuous development in the exploitation of traction power of draught animals was only one aspect of the increased use of power in agriculture, transport and manufacturing. But since it accounted for some 70 per cent of all power used (in Domesday England) it is worth considering (Langdon, 1986, p. 20). Windmills supplemented watermills in the mediaeval era but have often been thought of as a technological import from west Asia. Being different in design and efficiency from their alleged origin the European types can in fact be interpreted as emerging from the tradition of watermill construction and would consequently be of independent nature. Changes in waterwheels, sails and towers of windmills increased their efficiency gradually (Singer, 1956, pp. 617–18 and 623–8). With the invention of the crank, which converted rotary motion to reciprocal movements, a lot of new productive activities were opened to the penetration of water and windpower. At the end of the mediaeval era mills were found in tanning and fulling, in metallurgy as a prime mover of bellows and hammer-forges, in sawing, in pumping water from mines and fens, in irrigation, and, of course, in the traditional agricultural tasks such as grinding grain, pressing olives, etc.

The evolution of traction power was no less impressive. The general trend, although slow, was to increase the degree of utilization of the draught-power capacity of animals. It was accomplished by a series of changes in implements. Most important of these was the application of better harnesses and traces. The ancient harness with neck and body girths had the disadvantage of impeding the breathing of the animals if much draught power was needed. The neck girth pressed on the windpipe of the animal, more so for horses than for oxen it seems. Through some intermediary improvements the mediaeval collar was developed. In contrast to the girth, this rested on the shoulders of the animal thus avoiding the problem mentioned above (Jope, 1956, p. 554). The combined effect of improvements in harnesses and traces was tremendous. It contributed to an estimated three or fourfold increase of draught power. But again, this change did not occur in one stroke (Klemm, 1959, p. 80; White, 1978, p. 19). Similarly the use of nails can be seen as a more efficient way of attaching the horseshoe, but the shoe was known before mediaeval times. The introduction of double shafts (replacing single vehicle shafts) improved breaking power, and changes in vehicle design such as lighter spoked wheels had a similar effect – see Langdon (1986, ch. 1) for a recent survey of important technological changes in the use of draught animals.

Although individual animals were more adequately used in the late mediaeval era than in any previous period, the tandem harness was a considerable advance in the use of the combined power of draught animals. When harnessing the animals abreast, which was the traditional method, much energy was lost because the traction was performed at an angle to the direction of the vehicle or plough. There was close interdependence between the efficiency gains in draught power and the evolution of ploughing, as well as the development of vehicles for transport of goods and man. In the classic study of the harnessing of horses, Lefebvre de Noëtte estimates that the compound effect of improvements including the introduction of tandem-harnessing was a tenfold increase in traction power from teams from the early Middle Ages to the nineteenth century.[6]

There was also a change in the source of traction power in that the horse gradually replaced the ox. The horse was, however, expensive in its fodder requirements compared to the ox, which has made some historians doubt the alleged speed with which it became the primary draught animal. There are regional differences suggesting, for example, a more rapid penetration in France than in England. But not all types of agricultural production units were equally swift in adopting the horse. J. Langdon has advanced and defended the view that peasant households (in England) were quicker to adopt the horse than manorial units. The reason is supposed to be that oxen were fed by inferior fodder such as straw and hay with low or zero opportunity cost at manors. Peasant households, however, could not overlook the costs of straw and hay, making the fodder cost differential between horses and oxen smaller than usually believed (Langdon, 1982). There also seems to be an association between the introduction of the horse in agriculture with the transition to a three-course rotation system. Oats, which were the preferred fodder for horses in periods when they worked hard, could be sown as a spring crop in the three-course system. Oats became an important crop in the late mediaeval era which makes them an outstanding example of the selection of new and suitable crops complementary to the technological transformations just reviewed. Let us therefore turn to a discussion of that sequence.

The Crop-selection and Specialization Sequence

Direct genetic manipulation of plants and animals is a fairly recent phenomenon, but pre-industrial societies nevertheless made important contributions to the selection and improvement of crops and livestock. The spontaneous process whereby natural selection operates on chance variation has been supplemented by deliberate human selection. The emerging pattern was that of a greater variety of crops in cultivation and the seed varieties chosen were increasingly well adapted to the specific requirements of soil and climate (van Houtte, 1980; Abel, 1980b, pp. 539–40).

This is of course the simplest process of growth of knowledge in which impulses for change might sometimes be an *erreur heureux*, and sometimes growing demand for plants known in the environment, such as some of the industrial plants that were used in late mediaeval textile manufacturing. The point can also be illuminated by the introduction of oats, becoming – as was noted above – the 'fuel' for horses that replaced the oxen. Originally, oats appeared as a weed in wheat-growing, however. What made them attractive apart from their use as fodder and in the breweries was that they could be cultivated on meagre soils and in harsh climates.

The better match between crops and environment was assisted by the revival of trade, but there were also efforts made to ameliorate the quality of plants resulting in higher yields or improved products. Before cider became a popular drink at the end of the Middle Ages, attempts were made to use wild varieties of apples. Through selection and cultivation the properties of the fruits were gradually modified so that the sugar content of apples was increased and, as a consequence, the cider became more agreeable.

Trade stimulated the emerging pattern of regional specialization into predominantly wine-growing, grain-growing and livestock-farming areas. Demand and transport facilities, as well as a more profound knowledge about soils, climate and plants, were the prerequisite for such a change. Close to cities, gardening and industrial plants were cultivated on an unprecedented scale: flax and plants for dyeing in the textile trades, hops for the breweries, and a wide variety of legumes and fruits for the urban market.

The selection of higher-yielding crops is obviously a technological change and by and large we can expect that regional specialization led to productivity gains: each region exploited its comparative advantage in climate, soil and experience. The effects of the introduction of *new* goods – a more varied consumption pattern undoubtedly evolved in the late mediaeval era – is less easy to characterize. We can, however, view it as a quality change in food and shelter, and as such a gain in productivity. Not necessarily *more* food, in other words, but better or *preferable* food.

1.7 Conclusion

In this chapter, technological progress has been viewed as the spontaneous and endogenous outcome of production. Technological progress stimulated population growth in pre-industrial economies *and* was stimulated by population growth insofar as it increased the aggregate demand, which generated division of labour, regional specialization, and liberated the economy from some of the constraints imposed by indivisibilities in equipment and learning. The social conditions for technological progress – viewed as a slow process of accumulation of knowledge – have so far been abstracted, apart from a reference to a need for cultural continuity. In the next chapter I will discuss what is entailed in cultural continuity and investigate the social conditions for technological change in pre-industrial societies.

NOTES

1 One may argue that income is an inappropriate concept in an economy which is only partly monetarized. Income, however, is used as a synonym for output and product.
2 Technical, in Elster's terminology, is technological in the usage adopted in this book. In a recent survey and analysis of theories of technological change Elster adopts a less repudiating view as regards technical and technological change in pre-industrial economies (1983, pp. 131–7).
3 Technological change will typically reduce the input of labour and other resources per unit of food produced and extend the knowledge of the environment so that a greater part of the available biomass actually is exploited. Oysters and shellfish, for example, were added to the normal diet at a rather late date.
4 When a group of hunters kill the deer the number of animals will diminish, which may decrease the productivity of the territory for them and for all other hunters in that area. Insofar as that deterioration in productivity affects this particular group it will be taken into consideration by them, but the social cost (the decrease in productivity) that the group inflicts upon other groups will not be. This is the source of the over-exploitation of common property in the absence of institutions that regulate its use.
5 There are several informative surveys of mediaeval agriculture which pay due attention to technological conditions; see, for example, Daumas (1962), Kellenbenz (1980), Myrdal (1986), Parain (1979) and Singer (1956). P. Mane (1983) has provided admirable documentation of mediaeval technology as it has been represented in art.
6 More precisely the comparison is made between the maximum load permitted by the Theodocian code (438 AD) and observed capacity by the middle of the nineteenth century. See Langdon (1986, p. 8).

2 Individualism and the Elementary Social Conditions for Technological Change

2.1 Introduction

According to the stagnationist view, there was little technological change and economic progress in pre-industrial societies. This perspective is either justified by various *ad hoc* explanations or accounted for by the absence of efficient organization and adequate institutions.

Institutional failures are normally ascribed to the prevalence of common property and to incomplete property rights, which generate externalities. Externalities occur when the action of an individual (or a household or a group of individuals) affects the welfare of other individuals (households, groups). Irrespective of whether the external effects are beneficial or harmful they create an incentive problem because private and social costs (benefits) diverge. If some of the benefits of an individual's action accrue to others, the generator of that beneficial externality will not be sufficiently rewarded to undertake the effort and the volume of production will be inefficient. If, on the other hand, the externality is harmful this implies that the social cost fails to be adequately reflected in the private costs of the individual or household. In the first case, an institutional failure exists because there is no scheme that rewards the producer to extend the effort to a socially desirable level, and in the second case the problem is how to restrain the producer from a socially undesirable action.

While abstracting from transaction costs and costs of implementation and supervision of property rights, it has been suggested that externalities could be handled by negotiating compensations and subsidies among members of the society. This implies an extension and redefinition of property rights so that, for example, a generator of a beneficial externality could claim a subsidy from other members. In the case of a negative externality, conversely, members of society could claim compensation. More generally, it has been suggested that by internalizing externalities in private costs (benefits) the divergence between social and private costs (benefits) would disappear and – under perfect-competition assumptions and abstracting from transaction costs – efficient outcomes should therefore ensue.

The intellectual inspiration behind this approach derives from a growing body of theoretical inquiry called the property-rights paradigm that was sparked off by R. H. Coase's paper, 'The Problem of Social Cost' (1960).[1] That approach is quite rich in implications depending on whether transaction costs are dealt with or not, but a disproportionate effort within that approach has been geared towards establishing the case for gains in efficiency from introducing private property rights over scarce resources. In an early application of the hypothesis, H. Demsetz explained the evolution of private property rights over territory among Labrador Peninsula Indians (1967). Before the arrival of the French colonialists there were communal rights in the game. Hunting, however, introduces an externality in that the individual hunter passes over some of his costs to the other members in society: hunting reduces the number of available animals which will increase the costs for all hunters by forcing them to work harder in future. (Assuming quite realistically that the time required for a successful hunt is on average related to the number of animals available.) As long as resources were plentiful and used for local consumption only – and perhaps because of collective regulation of the hunting – there were no immediate problems in terms of scarcity. When colonialism provoked an increased demand for furs, the signs of over-exploitation of the game were acknowledged. Over-exploitation is explained by an institutional failure, in that the private loss amounting to the income foregone by not hunting a particular animal is considerably higher than the private loss in the gradual depletion of the stock, since most of these costs are shared by other members of the community. Demsetz argued that private property rights in the land and its game developed in response to this potential over-exploitation crisis and that this brought private and social costs closer to each other and induced a socially more efficient use of the scarce resource.

On a more general and ambitious level, the property-rights paradigm was first applied to economic history by D. North and R. P. Thomas who argued that the success of the Western world from the early modern period and onwards can be explained by institutional changes that made social and private costs (benefits) converge (1973, pp. 1–8). By rewarding and imputing costs more efficiently, individual effort was geared towards socially and privately desirable ends. Typical institutional changes in this direction, they argued, were the privatization of the common, the introduction of patent rights and the abolition of serfdom.

If the costs of implementation and supervision of property rights as well as transaction costs, i.e. costs of negotiating compensations and subsidies among members, are taken into account the theory becomes compatible with a large variety of institutional arrangements. This fact is acknowledged by North and Thomas who try to explain the prevalence of non-privatization of resources in pre-industrial economies by the high transaction and supervision costs: the gains from the establishment of property rights may have been fully matched by the costs associated with monitoring and negotiation. *A priori*, not much can be said about the relative weight of gains and costs and there is clearly no ground for an argument that private property rights are the only alternative to free access to a resource, or that private property rights are always superior to other arrangements. In fact one of its proponents among historians, C. J. Dahlman, has provided a property-rights analysis of collective property in land used for grazing, i.e. the common (1980, pp. 115–21). The Dahlman hypothesis, which is discussed at some length in section 2.4, states that the combination of a technological characteristic – economies of scale in grazing livestock – and high transaction and implementation costs associated with private property rights made collective property rights – i.e. some type of collective regulation of the access to a resource – a more efficient institutional setting.

Even if transaction and implementation costs associated with private property have been acknowledged in the property-rights approach, the heuristic power of the theory is greatly overstated by its partisans. Contrary to the claims made by the property-rights school, externalities are not always generated by defective property rights. An important source of externalities originates in the power which an individual will have in all situations where other members of the society will be worse off without that particular individual. Power, defined in this way, will exist in all economies that are not very large and it constitutes a major problem for all small societies.

When individuals have power in the sense that their withdrawal can exert harmful effects, then members of society can and must negotiate the conditions under which they participate and exchange goods and services. If members have access to (property rights in) different resources or personal capacities such as specialized occupational skills, the terms at which these resources or capacities are exchanged – i.e. the prices – will be a matter for bargaining. This problem has long been overlooked in traditional economic theory: the familiar results of efficiency (Pareto-optimality) have been proven under the assumption that prices are given and cannot be manipulated by the participants in the economy. In small economies this is manifestly untrue and it is strictly true only in infinitely large economies.[2]

Small economies are as a consequence bargaining economies in which many of the traditional concepts and results of economic analysis do not apply. In small economies, members have power and can generate externalities and it can be expected that the outcome of a bargain, that is the production and the distribution of the income, will vary from time to time. The resulting outcomes will depend on the configuration of strategies that members adopt, the credibility that members ascribe to their contenders' threats, the information they possess on the motives and strategies of others, aggressiveness and the skill performed in the bargaining process, etc. It is also true that these outcomes, in general, will not be Pareto-efficient.[3]

2.2 An Agnostic View on Social Institutions

These critical remarks invite us to take a more relativistic view of social institutions and property arrangements. The overall efficiency of private property rights and self-seeking individualism will be shown to be a misleadingly simplistic view. A major point of this chapter is to explain why relatively unrestrained individualism is a comparatively recent phenomenon and why social control of, and constraints on, individual rights have been such a dominant feature of pre-industrial societies; and, in fact, a necessary element in their ability to develop technologically.

Although this analysis stresses the historical relativity of social institutions, that does not imply that institutions are arbitrary. On the contrary, there is a distinct flavour of historical materialism in the general outline of the suggested explanation. It will be shown that factors of a technological or technologically derived nature create specific constraints and conditions for human interaction. Given these constraints a series of typical and often misinterpreted pre-industrial social institutions will be shown to be adequate (suitable) in the sense that they contribute to technological progress. These two propositions, the technological determinant of social institutions and the adequacy characteristics of stable social institutions, are weak versions of classical historical materialism.

G. A. Cohen (1978) has provided a rigorous elaboration of historical materialism in which he argues that historical materialism should be considered a functional explanation. For the purpose of the arguments in this chapter, the two relevant statements in historical materialism are (1) that the level of development of the productive forces (which is a concept that entails technology as understood in this book) explains the nature of its economic structure (roughly, rights and economic institutions in the terminology followed in this book), and (2) that the economic structure is responsible for the development of the productive forces (roughly, technological progress). These two arguments will be interpreted as a claim that the economic structure exists because it contributes to technological progress given the prevailing character of technology. It is not sufficient to show that the economic institutions actually have such an effect to assert that they owe their existence to it. There must exist a mechanism that shows how their having such an effect explains their existence (Cohen, 1978, p. 271, cf. also the exchange between Cohen, 1980, and Elster, 1980). In the final section of this chapter this problem is discussed again but without, I am afraid to say, any conclusive and strong results. So until then the aim is to show that pre-industrial social institutions in fact can be related to the existing level and nature of technology and that they have contributed to technological progress. These statements are equally controversial so they are worth considering in their own right, besides being necessary, albeit insufficient, for a historical-materialist theory of institutions and institutional change.

Social institutions entail property arrangements but include other arrangements that regulate human behaviour and interaction, such as morals, ideology and social conventions. Such institutions will be called adequate insofar as they contribute to technological progress. As argued in chapter 1 the general requirement for technological progress is social continuity. Let us therefore first investigate the precise content of that requirement.

Social continuity presupposes a relatively stable community where technological knowledge is accumulated and transferred from generation to generation. The community must have a certain stability in its membership, which in most cases implies some degree of social order or social cohesion. This is a precondition for members being able to communicate and share the knowledge which has been gained. It is also obvious from the argument in the preceding chapter that technological progress will be enhanced if the community that shares the endogenously generated knowledge can be enlarged either by increasing its membership or by entering into contact with other communities in an orderly way – rather than in a belligerent manner. Endogenous technological progress is related to the number of productive operations performed where the latter in turn is determined by the size of the community and the level of per-capita income. The size of a community or groups of communities cannot be increased without orderly forms of interaction of members. So, by and large, social continuity presupposes institutions (rights, conventions, rules of conduct) that generate the conditions for members to engage into permanent and peaceful relations.

It may be argued that if members of a society realize the mutual advantages they have in establishing a cultural continuity there is no need for elaborate social institutions at all. It is true that some social conventions, such as, for example, language, seem to be self-enforcing. Men contribute to the sophistication of language because there is a mutual advantage in understanding and being understood. Following that logic it could be argued that the advantages of social order and continuity are also self-enforcing insofar as there are mutual advantages in such a development. But, for reasons that are discussed at length below, this is not generally the case.

It will be argued that typical and widely spread social institutions in pre-industrial economies are dependent on three sets of technological characteristics.

1   Let us first investigate the very level of technological knowledge. At a low technological level concerning production, conservation of products and transport, two important problems will arise with implications for the social organization of production.

Firstly, the economy will be comparatively small, i.e. it will have a limited number of members, for two reasons. Primarily because a low technological level implies that the means of transport are not developed enough to permit trade of goods at a vast scale over wide distances. Moreover, although important land-saving innovations occur in pre-industrial economies the technology will remain fairly land-extensive. That also implies that the economy will be small because the density of population is limited by the high input of land needed per household. The social consequences of the size of the economy are presented under point 2 below.

Secondly, a low level of technological sophistication also means that the production process is susceptible to severe disruptions by exogenous shocks. Most important among these uncontrollable disruptions are climatological factors, but one should also mention diseases of man, animals and plants. There is in other words an undeveloped control of nature. Assuming that these random shocks have a zero and independent mean the effect will disappear in large samples of productive operations. An equally undeveloped technology in the conservation of food will however make an intertemporal redistribution of products within the household unattainable. This technologically derived predicament is probably the most basic rationale for the existence of groups or bands of households providing an element of risk-sharing and redistribution.[4]

But risk-sharing, although it is an advantage to all members under the specified technological circumstances, also causes a difficult problem of social organization of production. When there is pooling of risks within a group the welfare of an individual or household will depend not only on its own effort but on a joint product as well. The production of a household not only satisfies its own needs but the need for a surplus that can be redistributed in case there is a subsistence crisis. If each member (household) produces just to the point at which the private utility or satisfaction equals the disutility (or dissatisfaction) of labour this will be socially inefficient. This inefficiency arises from the fact that there is a positive externality present in that the rest of the society would gain from a higher productive effort since it would increase the total product and consequently their capacity to resist risks. It can be called a partnership externality (see Radner, 1986, pp. 16–18).

The particular household we have been describing would be better off if all other members increased their effort by a small amount. Thus, the inefficiency cannot be overcome unless some accord on desired levels of effort can be agreed upon and monitored. Especially in small groups, when the dependence on a joint product is relatively more important, this externality may prove to be decisive. This need for risk-sharing may explain the prevailing high level of social control of individual effort in small economies which have a high incidence of exogenous disruptions of production and a low control over nature. As a consequence the evolution of comparatively unrestrained individualism is positively related to the size of the economy and to the level of technology. More specifically it is argued in section 2.4 that open-field agriculture with scattered fields – so confusing for many historians and described as hopelessly inefficient by others – is a device for monitoring individual work effort whereby the volume of the joint product can be controlled.

2 Another technological characteristic of paramount importance is the scale properties of the economy. Division of labour generates productivity increases and is, as shown in the preceding chapter, intimately tied to the scale at which the economy operates – 'the extent of the market', as Adam Smith preferred to call it. Bringing people together in stable economic relationships of division of labour is therefore susceptible to economies of scale in production. Scale effects are present in risk-sharing as well (see above), and to differentiate that effect from scale effects in production it will be called 'economies of size'.

The advantages with barter within and between communities are also a perennial source of instability and strife. This is so because the positive scale effects gained in production and economies of size in risk-sharing will give any individual the bargaining power that was discussed in section 2.1. By threats of withdrawal from a particular community, any member or coalition of members can put the rest of the society in a situation in which it is worse off. It will be argued that this peculiarity of small economies is one important determinant of social arrangements.

Division of labour also involves both training and investment in tools that exhibit indivisibility characteristics, i.e. a minimum of training effort or investment. It can reasonably be expected that individuals will be reluctant to engage in that type of investment unless there is some certainty about the possible proceeds of such an endeavour. But, as has been pointed out repeatedly, the bargaining process is costly and the outcomes of negotiations are difficult to predict. This argument and the one referred to in the preceding paragraph are developed further in section 2.5 in which the guild system is interpreted as a framework for bargaining giving a higher degree of certainty and predictability to outcomes, as well as diminishing the transaction costs in negotiations.

3 Finally, there are the problems related to the characteristics of goods and services. Indivisibility is one such non-trivial characteristic, which has already been discussed above. More interesting is the fact that some goods and services, so-called public goods, have a non-exclusive nature. The typical characteristic is that if it is produced it will be available to all. The consumption of a public good by one member will not diminish its availability to others, in contrast to private goods, such as bread. The typical case is security. If there is an efficient production of security, everybody in that area will be protected, irrespective of whether they have contributed to the production or not. As a consequence, a so-called free-rider problem may arise. You are a free-rider when you use the public good you have not contributed to. If all or many behave in this way or understate their true preferences for the public good – economic theory suggests that this does indeed happen – the production of the good will be insufficient. Some sort of social organization which helps to reveal the true preferences and control the contribution to the public good is therefore called for.

Also, in case one expects less self-interested and more altruistic behaviour, such as a principle that one is willing to contribute to a public good if others do so, it can easily be seen that this presupposes a considerable co-ordination of individual action. Some rules of conduct or morals do have the typical traits of public goods but can exist seemingly without support from any explicit social organization. Morals, usually interpreted as commands, may therefore be understood as information about the expected behaviour of others and may – interpreted in this way – serve to give an individual assurance concerning the behaviour of others. How this type of institution has emerged is not easy to say but it may be related to recurrent situations in which the shortsighted free-rider behaviour has proved to be disastrous in its social consequences. Rules of conduct and morals interpreted in this manner can be expected to be unstable, however. If someone starts to break them the tacit understanding of how people can be expected to behave disappears.

The problem of public goods is intimately related to the development of authority and the state in societies. This is such a vast field that it cannot be adequately treated within the framework of this book, the scope of which is restricted to an analysis of economic institutions.[5]

The arguments in this section can be summed up in the proposition that the very reasons for the search for community – self-interested or not – creates a need for elaborate social institutions to make that community viable. There are at least four reasons: (1) problems of risk-sharing which in fact is a problem of how to monitor contributions to a joint output; (2) the externalities that emerge just because the economy is small which creates bargaining problems much the same as those generated by (3) scale effects of production; and finally there are (4) the problems of public goods.

2.3 Feud and the Rituals of Exchange

Much of the discussion so far has been based on an assumption of man as a self-seeking and self-interested, and sometimes narrow-minded, creature. Is not man sometimes capable of exhibiting a more altruistic behaviour, or one that is based on a little more foresight? He certainly is. The stability of spontaneous rules of conduct in spite of their public-goods nature is evidence of this. The fact that men engage in collective action – in spite of predictable costs and uncertain rewards – in revolutions and reform movements provides additional, historically important evidence.

On the other hand, it is also clear that man often adopts a self-interested type of behaviour, more often in the economic domain than in social affairs, perhaps. There is a strong tradition within economic history and economic anthropology which argues that self-interested and rational economic behaviour is a recent phenomenon resulting from the unprecedented growth of market transactions in recent centuries; it is not – in this view – a general characteristic which can be observed throughout history. In pre-industrial economies, so the argument goes, exchange was deeply embedded in social customs and rituals taking the form of gifts and carrying an obligation of reciprocity. This position – known as the substantivist school in anthropology – had an influential mentor in Karl Polanyi and his *The Great Transformation* (1944) and it has obvious intellectual connections with institutionalism and the German historical school in economics. Although the thinking of Polanyi forms a living heritage in economic anthropology there has also been constant opposition from a so-called formalist tradition trying to interpret seemingly non-economic behaviour as economically rational. A typical case is to look at the gift as a form of insurance. There are recent studies that make these arguments in a rigorous way in which the self-interested nature of the gift is shown to be its reciprocity.[6]

Although this seems a valid point, the formalist approach and its forerunners do not give a complete account of the roots of the ritualized exchange. It is suggested here that ritualization of exchange is a way of hampering conflicts which arise when self-interested individuals in small economies get involved in trade and barter. As will be recalled from section 2.2, the small size gives members bargaining power which, under conditions of unrestrained self-interested behaviour, can prove socially disintegrative and unsuitable for social continuity. Rituals are but social conventions prescribing adequate behaviour. The anthropological literature is full of accounts of endemic conflict within pre-industrial economies but I will turn to a historical source that better than anything else tells a similar story: the Icelandic sagas.

Historians may raise some doubts about the usefulness of the sagas as an historical source. But what is at stake here is not the authenticity of the individuals or actual feuds described in the sagas. That does not matter much. It is, however, reasonable to interpret them as concerned with real and recurrent problems facing societies of this type.

In most sagas there are long chains of feuds between individuals, households and chieftains. In the process of the resolution of these feuds there are attempts at forming coalitions arranged through travels around the island by those immediately concerned. Frequent intervention of arbiters also occur. The conflicts which the feuds focus on are often material, but some also concern other sorts of relations, such as insults and seduction. The material conflicts emerge because of difficulties in reaching agreement about the distribution of land, inheritance, dowries, the management of shared property, the terms of exchange of goods and gifts, and, finally, theft and killings. The patterns of bargaining and resolution are also manifold. Sometimes they only involve the parties concerned. In other cases they are settled through collective bodies such as the Althing, the congregation of free men, and through the appointment of a good man.

It is a characteristic feature of agreements in the sagas that they are inconclusive even when a resolution of the immediate or initial conflict is reached. This feature is a typical characteristic of bargaining situations in that one party often believes that it can do better in a new settlement.

There is also the more obvious case of continued and often escalating conflict when one party rejects an offer. As a contrast to these socially disintegrating and often violent conflicts there are of course the peaceful and terminated conflicts.

Njal's Saga (1986: Penguin Classics) gives ample examples of all types of conflicts and their resolution. It also provides striking examples of the morals proclaimed by the sagas: *mutually advantageous relations between men presuppose mutual restraints in individual behaviour.* Gunnar from Hlidarend and Njal both incorporate these characteristics. Njal 'was a gentle man of great integrity ... his advice was sound and benevolent...'; Gunnar was 'extremely well-bred, fearless, generous, and even-tempered, faithful to his friends...' (pp. 73–5). Since they both have these characteristics they succeeded in peacefully solving the many and often severe conflicts thrown upon them by their wives. As a contrast to the resolution of conflict between Njal and Gunnar, often involving the transfer of vast amounts of valuables, there is the escalating and uncontrollable conflict caused by the fact that one of the parties and sometimes both parties lacked restraint. That type of man could be described as 'arrogant and brutal', 'aggressive and overbearing' (e.g. p. 140). Gunnar is involved in a chain of escalating feuds with people of that kind – in which Njal supports him. The feuds eventually cause violence from his side and finally his own death. But it is clear from the narrative that Gunnar tries hard to evade the escalation by offering generous terms of settlement which are rejected by his more aggressive opponents (e.g. pp. 119, 126). The chain of feuds starts when a farmer refuses to sell hay to Gunnar, while Gunnar characteristically accepts the offer to buy a slave from that same farmer well supplied with hay. 'It is mainly to share', as Njal puts it in another context, but this moral is not applied by all. The slave turns out to be not only defective but also a cause of the continuation of the feud. Facing ruthless men there is no peaceful solution at hand and finally but reluctantly Gunnar has to adopt the same strategy as his antagonists. So great is the moral resentment against the ruthless and overly ambitious man that there is a specific term for it (Byock, 1982, pp. 29–30). In fact if a man is pressing a claim with 'greed and aggression' then in the eyes of the saga narrator this forms a legitimate basis for not accepting an agreement (e.g. pp. 53–4).

It has been argued that restraints on individualism is a permanent concern in small pre-industrial economies because of the power each member possesses. In the following sections of this chapter a series of institutions are discussed as institutional solutions to the problem of order and cultural continuity in such societies.

2.4 Scattered and Open Fields

While manorial production represented a transitionary phase of mediaeval agriculture there was considerable stability in the institutional organization of the peasant village. Throughout Europe, although not everywhere, a system prevailed that was not only an intermixture of private and collective property rights but which was also associated with a peculiar 'layout' of the privately owned land. The waste or the common was used for grazing livestock and collective property. Households claimed private property rights to the arable land, however. But the typical household had its property scattered over the whole arable area of the village instead of having it consolidated into larger units. This arrangement imposed a relatively rigid set of rules on households on what to grow, how to work the land, when to sow and weed, plough and manure, and many critics have seen this rigidity as the cause of the inherent conservatism of pre-industrial agriculture. Property rights and individual effort were circumscribed by custom which, it was argued, was detrimental to economic development; the system was 'absurdly uneconomical' to use Seebohm's famous indictment (1896, p. 15).

The prevalence of this agrarian system – and there is evidence of similar although not identical arrangements outside Europe, in Japan, Africa and among New World Indians – has stimulated historians to take a more favourable attitude. This concern has also been shared by critics of the system. Some have tried to uncover the origin of the open-field system with its scattered or subdivided holdings, but explaining a plausible origin is not necessarily an explanation of its stability. Others have tried to find the rationale for its existence, implicitly (and mistakenly) arguing that if there is some rationale for a feature that will explain its existence.

In his recent survey on the historiography of the subdivided field, Robert Dodghson identifies five different explanations advanced in the last 100 years (1981). To these five, a sixth distinct hypothesis has recently been suggested by C. J. Dahlman (1980), a seventh by S. Fenoaltea (1976) and, as if that is not enough – given the fact that empirical evidence fails conclusively to discriminate against the existing ones (Yelling, 1982) – an eighth is suggested here.

There are two specific explanations which concentrate on the origin of the system rather than its function. When land was colonized and added to the arable land already at hand it usually required a collective effort. Therefore, the argument goes, it was a convenient way to distribute the newly acquired land by partitioning it among the households of the village. The other explanation points to partible inheritance as the main cause. Although both these explanations may have some value as accounts of the origin they do not really address the problem of why subdivision persisted. If there were no advantages with the system we would expect households to consolidate holdings by means of land markets or exchange of strips.

Among explanations that try to identify a beneficial effect of the system the classical one is associated with Seebohm. It states that the subdivision of holdings in long strips on an open field had to do with the joint character of the ploughing in which each member of a plough team held strips of land representing one day's ploughing. Although the argument has since been put in a less rigid form it relies on the technological characteristics of ploughing as a determinant factor. It is clear that ploughing involved indivisibilities and therefore made joint effort necessary, but there is no proof that this made necessary the specific layout of the fields.

In another early attempt to explain the persistence of subdivision the Russian economic historian Vinogradoff (e.g. 1892) related it to the egalitarian nature of the village facing the problem that available land was of different quality. If households should have (approximately) equal shares in land of varying qualities, subdivision was a feasible solution. This perspective ignores the actual social differentiation in the mediaeval village. Nevertheless, it may be an important argument if the egalitarian striving within the village could be explained. That problem is discussed further below.

Recent attempts to provide explanations of the persistence of the scattered-field structure are explicitly based on economic theory. D. McCloskey (e.g. 1976, 1984) has provided a rigorous formulation of a longstanding interpretation of it as insurance against risk. If damages induced by nature or animals are stochastically located over the whole village arable, then it would be advantageous for each household to scatter the ownership over the area as a sort of risk-spreading. With scattering a household owns a 'portfolio' of strips some of which attract high risks and others low risks. It is, borrowing a phrase from McCloskey, 'a diversification against localized disasters'. Consolidation would mean that a household was susceptible to the risk of making great losses. Disturbances affecting aggregate production, i.e. the whole village, will not be spread, however, since the total arable is not affected but only its distribution. Scattering, it is argued, involves a cost in production foregone because it is inefficient. This may still make it an attractive solution if alternative means of insurance were even more costly in terms of production foregone. Most critics doubt this, pointing at other types of insurance: redistribution within the village, storage, credit markets and gift exchange (Dahlman, 1980, pp. 60–3; Fenoaltea, 1976, 1977). Furthermore in the lord–peasant contract there was an implicit responsibility of the lord towards the peasant in hard times, which can be seen as a type of insurance.

Dahlman's own contribution relates the scattered fields to the dual nature of agrarian production, being both arable and pasture. There were obvious economies of scale in grazing livestock and the scattered fields had essentially the function of enforcing households to participate in communal grazing. In Dahlman's own words the scattering decreased both the 'hold out power of any one farmer as against the collective... and ... the net benefit of separating himself from the collective ...' (1980, p. 125). Although it may seem odd to relate the organizational structure of one activity – the arable production – to reap some gains in another – i.e. the grazing of livestock there – Dahlman's hypothesis claims to provide a solution to a problem which none of the other theories have been able to solve in a reasonable way: the explanation of the enclosures. Dahlman relates the enclosures to the increased specialization that dissolves the duality in agrarian production. With villages specializing either in arable or livestock production there is no need for an arrangement exclusively related to the existence of that duality.

Fenoaltea (1977, but see also the exchange with McCloskey, 1977) has provided an interesting hypothesis in that it departs from the conventional wisdom that portrays scattering as costly and inefficient, whatever its other merits may have been. Scattering is efficient, Fenoaltea argues, because households may use its scarce labour over a longer agricultural cycle without recourse to the labour market. With consolidated holdings a single household would not be able to plant and harvest the whole area when it was optimal to do so. Scattering implies land of varying quality and the optimal working period is prolonged for major agricultural activities. The weakness of the argument is that it does not demonstrate that the gains through scattering were significant, and it overlooks the prevalence of hired labour in peasant households.

This survey cannot do justice to the subtleties and sophistication of the different theories but it provides a background for the evaluation of an alternative suggestion that entails some of the important findings of the other approaches.

It is clear that the Vinogradoff hypothesis would be considerably strengthened if some endogenous explanation could be given for the egalitarian bias in the village. Even if that bias did not prevent social differentiation it was nonetheless a force countervailing inequality. An egalitarian bias can in fact be interpreted as a self-interested response to the conditions of production in an agrarian village. Bearing in mind that (1) there were significant economies of size in a stable village organization because of the team character of certain tasks such as ploughing and protection, (2) size itself is a risk-absorbing feature, (3) a relatively homogeneous village could do better in the negotiations with the landlord, and (4) the village, because of undeveloped technology, still had a high self-sufficiency, then it would be rational to prevent the pauperization and drop-out of a significant part of the members of the community.

The capacity to overcome natural and climatological disturbances affecting a single household or the whole village is evidently dependent on the level of the aggregate product in the village or in the community practising redistribution, gift exchange and/or establishing credit-debtor relationships. This can be interpreted as a dependence on a joint product and a partnership externality will consequently arise (cf. above, pp. 38–9). Unless some sort of monitoring of work effort is agreed upon the production will be inefficiently low if producers are motivated only by their own satisfaction and if they neglect the positive externality related to their work effort.

In this interpretation, the essential characteristic of the open-field system and its scattering of property is that it enforces a production process that ensures a collective monitoring of work effort and aggregate production by means of deciding when, what and how to produce. Critics of the system often deplore this since, it is argued, it hampers individual ingenuity. The interpretation suggested here goes to the heart of the matter by giving a meaningful function to the very supervision and regulation of individual effort, which in many other explanations has been seen as an unfortunate consequence tolerable only because it is outweighed by some advantageous aspect of the system.

It can now also be seen that this interpretation is not necessarily opposed to the ones proposed by McCloskey and Fenoaltea. If Fenoaltea is right in claiming that scattering ensures an optimal allocation of labour over the agrarian cycle this will only strengthen the case for scattering having also a risk-spreading capacity, since no costs – contrary to the belief of McCloskey – are attached to it. But this device will not exhaust the need for insurance because households are only partly insured against 'localized disasters' and will not be insured against risks affecting the whole village.

This perspective stresses the monitoring of a joint product as an insurance; the dissolution of the open-field system and the enclosures must be related to new ways of coping with risks. The commercialization of agriculture is an important factor because it reduces the role of village self-sufficiency and increases the dependence on the outside economy. This process accentuated the social differentiation already existing so that a significant part of the agrarian population found the balance of costs and benefits (protection, insurance and security) provided by the village turning against open-field agriculture. The supervision and control of fellow village members is increasingly replaced by the discipline and advantages of the market.

This is not to say that commercialization made the inherent risks in agriculture disappear but agriculture had reached a higher level of productivity and other means of coping with risks, specifically credit markets, had become less imperfect. Well-to-do peasants felt, rightly it seems, that they could not fully exploit the market opportunities because they had to succumb to the decisions of a collective body. Unsurprisingly, the enclosures caused severe social conflicts and the very inertia in the process is an indication that the commercialization and the new dependence on the market was not equally advantageous for all. For the less fortunate peasants the advantages with the increasing dependence on the market were few and the disadvantages were great because they were not productive and wealthy enough to keep stocks. When village solidarity disintegrated they found themselves victims of usury rents and many entered a vicious circle of poverty in contrast to the thriving agrarian entrepreneurs in their vicinity. The poor needed the wealthy but the well-to-do no longer needed the poor.[7]

2.5 Guilds and Competition

The nature and effects of guilds – i.e. the corporate organization of urban production along lines of skill and occupation – have always aroused passionate controversy among historians and economists. Adam Smith, who as a contemporary witnessed the final phase of their decline, did not approve of their interference in the market and his spirit is present in modern appraisals of the guild system. In the modern discussion, G. Mickwitz's work (1936) is often cited as a typical exposition of the view which emphasizes the detrimental effects of the guild structure. A guild is a cartel, Mickwitz says, that raises prices above their competitive level by restricting the entry into the occupation and by limiting production (1936, see, for example, pp. 13–14, 66 and de Roover, 1958 for a similar point). There is an elementary objection against this argument that points out that it is an illegitimate generalization of a partial analysis. The mediaeval and early modern urban economy was not composed of one but of many guilds that negotiated with each other under the guidance of city authorities. In such negotiations the outcome cannot be grasped by the cartel metaphor which implicitly assumes a duality between a cartel and a non-cartelized sector, typically as the well-organized producers against the unorganized consumers. It is, however, not possible to make a distinction between producers and consumers because most households were members of guilds as producers and simultaneously consumers of goods manufactured by members of other guilds. Adam Smith, incidentally, made the mutual dependence of the guilds very clear, accentuating that the adverse effects of guilds rested in their superior negotiating strength *vis-à-vis* the rural sector.

The thesis expounded by Mickwitz has not been generally accepted by historians. This is not primarily because of its theoretical weakness but because it is believed to be at odds with historical evidence. S. Thrupp expresses a widely held view among the historians when she suggests that '... in any mediaeval town, at any given time the various guilds would have been strung out along the scale of economic power, with most of them bunched in positions in which they could have exercised little or no influence on selling-prices in local trade' (1963, p. 263).

Although Thrupp and Mickwitz reach opposing conclusions they seem to share a conviction that in the absence of guilds the typical urban mediaeval economy would have generated a competitive equilibrium. Thrupp in fact suggests that the guilds did not influence market prices much while Mickwitz deplores their doing so. Both err, however, because they take the view that equilibrium prices – which by definition are such that no single agent or coalition of agents can influence them – were possible in small urban economies of the mediaeval period. The nature of these economies makes a spontaneous evolution of stable prices that producers had to take as given very unlikely indeed. This was so simply because members had market power and in most cases a number of them believed that they could be better off if permitted a new round of price determination.

Given the premise – discussed more directly in section 2.2 – that all small economies are bargaining economies, the institutional structure of the urban economy in mediaeval and early modern times becomes intelligible. When prices are not exogenously given, members of the economy enter into head-to-head negotiations to determine them. Bargaining situations have, however, a series of characteristics that are undesirable in the sense that they may cause strife and friction among members of the economy and the bargaining process is itself associated with high transaction costs. Outcomes will depend – among other things – on the distribution of resources among members, negotiating skill, aggressiveness, bluff and ruthlessness or lack of it, credibility which may depend on previous strategies and outcomes, and so on. The Schoolmen of the mediaeval Church were certainly aware of the nature of bargaining, and its potential destabilizing effects were a matter of public concern (cf., for example, Brants, 1895, p. 200). Jacques de Vitry expressed a conviction that may have been shared by many of his thirteenth-century contemporaries when he said: 'Cheating, fraud, lying, perjury, circumvention and deception roam through all market places' (quoted in Baldwin, 1959, p. 67).

For these reasons the outcome – for example, the price of a good – will vary from negotiation to negotiation, and people normally express moral resentment against the fact that outcomes vary for no other reasons than those arising from the vicissitudes of the bargaining process (see de Roover's discussion on San Bernadino of Siena on this matter, 1967, pp. 12–22).

Although the Schoolmen were against the arbitrary element they did not oppose price fluctuations that had their background in demand-and-supply conditions, except in situations where supply was affected by severe harvest failures and demand by entitlement crisis. There are in fact some rudimentary traces of a utility theory of prices in the teaching of some scholastics (Gordon, 1975, pp. 227–8; Baldwin, 1959, pp. 73–4). Likewise the often misunderstood theory of the just price is basically related to the creation of ideal conditions for bargaining in which all arbitrary or abusive factors have been neutralized, such as assymmetry of information and power, coercion and fraud. To suggest, as de Roover does, that the just price is simply the prevailing market price is to neglect the concern the Schoolmen showed for the establishment of an ideal setting for bargaining. To accomplish a just price it was often seen as necessary, especially in cases where transactions did not occur very often, to invoke the judgement of a disinterested third party. This involved a price-fixing procedure described by a contemporary as 'a sovereign tribunal of arbitration where all the rights of all the weak and all the strong economic factors are taken into account' (quoted by O'Brien, 1920, p. 116), which resulted in a so-called *communis estimatio*.

We can safely assume that the Schoolmen represented a coherent view about some prevailing problems associated with trade and barter. Their concern arose out of an interest in social stability which they rightly believed was dependent on fair trade as opposed to abuse. Risk-taking, for example, became a legitimate source of income and there was an attempt to distinguish clearly between usury and acceptable interest. The Church was on the whole less rigid in its relation to the economic community than commonly believed. J. Le Goff makes the provocative point that purgatory was invented just to house the growing numbers of bankers and usurers that emerged with the intensified commerce in the dynamic phase of the mediaeval economy (1986). This may at first sight seem to be a modest improvement, but, bearing in mind that purgatory was a preparation for heaven, it probably gave great relief to those expecting to end up in hell, and the Church was quick to exploit this economically. If the Schoolmen were correct in their diagnosis of the problems that were associated with commerce in mediaeval and early modern urban economies, in what way did the guild system respond to them?

A guild provided members with rights and obligations. A guild had a monopoly or a semi-monopoly in manufacturing and trade in its products within the city. But these advantages did not go as far as a right to determine the size of the guild nor the prices of goods. These were matters of bargaining between the guilds and city authorities. The obligations put on the guilds arose from a desire to establish and supervise rules of conduct, for example to combat fraudulent behaviour among members of the guild. The city authorities delegated jurisdiction over such matters to the guild. A guild was also in charge of quality and quantity standardization which obviously decreased transaction costs. The reason why the jurisdiction over members was delegated to the guild had presumably to do with the fact that the guild had better information about the specific conditions relating to each product.

It is not the case that the guild system imposed a bargaining structure upon an economy that could do without bargaining. The guild structure only provided an institutionalization of the bargaining process which is necessarily associated with an economy of this type. There was no way in which the city authorities could prevent collusion between members in an economy as small as the typical mediaeval city. Through the guild system, collusion was institutionalized into co-operation based on a balance of rights and obligations more easily manageable for the city authorities precisely because it was not secret. It also contributed to diminished transaction costs since prices were regulated and quality standards upheld. This made the outcome of economic activities more predictable and acted as a stimulus to producers to invest in skills and equipment necessary for the attainment of specialization.

Some of the regulations typical of the guild system can be interpreted as a means of creating an even distribution of bargaining power. In some cases they prevented some members from gaining considerable power, which we would expect them to exploit and which contemporaries knew them to exploit. Given the fact that economies of this type were susceptible to exogenous shocks and disturbances in supply, the volume of resources a single member had in his command was directly linked to the bargaining power. For that reason there were prohibitions against 'engrossing' and 'forestalling', that is any type of hoarding for speculative reasons. A single member could not monopolize a source of supply but was forced to resell raw materials to other members at the price of purchase (Gross, 1890, p. 49). The general animosity towards usury can also be seen as an awareness that usury rents typically appear in imperfect markets where the debtor has a very weak bargaining position. When credit became part of commerce, rather than a way to bolster up a sudden loss of income, debtors and creditors were on a more equal footing and such instruments of credit developed rapidly. For some time rent was, however, disguised as a differential in exchange rates of, say, Genoese lire in Genoa and, for example, Palermo (see, for example, Jacques Heers (1973, pp. 238–9) for a description of how this credit institution worked).

The authorities relied on their control of the guilds to manage entitlement crises that emerged from adverse supply-and-demand conditions. In such situations price controls and attempts to increase or ration supply were tried. The balance which guilds enjoyed between advantages and obligations contributed to make them co-operate with the local power. Given the susceptibility of the mediaeval economy to exogenous shocks, it is not surprising to find institutions that we normally associate with conditions imposed only by war in the modern era.

It is also possible to interpret the guild structure as contributing to the insurance of the members. A sudden decrease in demand for the products of a guild is evenly spread within the guild because of the limited role played by competitive price offers. And in periods of supply restraints, the rules did not permit any member to hoard raw materials; they had to be made available to other members in the guild. Apart from this type of implicit insurance the guild normally helped members in case of sickness and death.

The basic contribution of the guild system was to change the nexus of bargaining from head-to-head encounters to negotiations between organizations representing the trading partners as well as, in most cases, a third party representing the public interest. Whether this was an efficient arrangement is not easy to say because we do not have straightforward efficiency results for bargaining outcomes. In any case, it would be illegitimate to compare the actual outcome with an ideal and unattainable market outcome. The system seemed to be adequate and even progressive in the sense that it promoted the hitherto unseen specialization in crafts and led to the perfection of skills which occurred in the mediaeval and early modern period.

2.6 The Social Acceptance of Individualism

Most theories of economic development ascribe a fair deal of importance to acquisitive and self-seeking individualism, whilst neglecting the importance of collective action and political institutions. There is also general agreement in the literature that the first signs of a new and more forceful individualism in the economic arena can be noticed in the cities of advanced regions of Europe by the fourteenth and fifteenth centuries and that individualism became increasingly important in the following centuries. Some even argue that individualism penetrated the quiet life of peasant communities in the mediaeval period (Macfarlane, 1978).

According to the property-rights approach, individualism was stimulated by the extension and redefinition of rights so that individual effort was more adequately rewarded than before. Institutional innovations in banking and insurance certainly facilitated trade and production beginning in the wake of the early modern period. But it must be stressed that the mediaeval economy already had a developed system of property rights both individual and collective. Peasants, it is true, succeeded in negotiating well-defined rights and there was an upsurge of market transactions concerning land and property. Although rights and incentives are intimately related – as is discussed in detail in chapter 3 – it will be argued that the preconditions for the evolution of self-seeking individualism primarily lie elsewhere.

There is a less cogent but no less relevant account stressing increased opportunities as a driving force behind the awakening of individualism. But the most controversial, the most discussed, and probably the least relevant, is of course the Weberian story. Contrary to popular belief Weber did not argue that the Reformation (or more precisely the Calvinist tendency) generated acquisitive individualism. He believed it to be a fairly constant human behavioural trait. The contribution of Protestantism was that it geared that individualism to a rational pursuit of worldly goals: the acquisition of material wealth was turned into a calling. R. H. Tawney, being rather critical of Weber (cf. Tawney, 1926, pp. 212–13, 319–21), aptly and poetically describes the transformation discussed by Weber as a change in 'moral standards which converted a natural frailty into an ornament of the spirit, and canonized as the economic virtues habits which in earlier ages had been denounced as vices' (from Tawney's introduction to the English translation of *Die protestantische Ethik*, Weber, 1958, p. 2).

Whatever the merits of these approaches – the Weberian thesis and its many critiques have developed into an academic subspeciality – they tend to evade the important question of why individualism, being such a potent force, did not develop before. That very question can however be answered within the framework expounded in this chapter. One characteristic of economies at a low level of technological development and per-capita income is their embryonic division of labour. This was discussed extensively in chapter 1 in which it was also pointed out that the combination of a rudimentary division of labour, low income and undeveloped transport technology necessarily constrained the size of these economies, i.e. the number of members remained relatively small. Even if there are examples of political units of considerable size in pre-industrial times, such as the Roman Empire and the Chinese dynasties, which also stimulated interregional trade and development, most economic transactions were restricted to small units, such as the village or the city with its region. The small size of the economy is an independent source of externalities: each member has considerable power in the sense that his or her presence in the economy affects the well-being of others. The power of each individual represents a crucial difference compared to a large economy in which a single individual does not really count: the well-being of others will not be affected by the withdrawal of that particular individual. The problem does not – as was pointed out above – arise because of defective property rights but because of the technologically determined size of the economy.

When per-capita income is increasing and the division of labour gains momentum, producers become involved in larger economic networks. The village economy with its partnership structure is increasingly superseded by production for an anonymous market which is somewhat unpredictable in its movement and outside the control of a single producer or buyer. The same tendencies are even more accentuated in urban production in the early modern period. The economies become larger or at least large enough so that producers cannot collude as easily as before and exert the same influence because of national integration of markets and international trade. Those who engage in markets with many participants are indifferent as to what a single producer does. This leads us to the paradox of individualism: *individualism becomes important as a social force when the economy is large enough to make each individual relatively unimportant.*

The paradox relies on the general argument pursued in this chapter that small economies typically are bargaining economies in which conflicts over exchange can escalate in the absence of rules of conduct that restrain aggressive and unbound individualism. Furthermore, partnership-like production so dominant in pre-industrial agriculture develops a need to monitor the efforts of others. In large economies, tolerance towards the individual, the acceptance of self-fulfilment and more advanced rights – in a word, individual freedom – are accepted because the potentially harmful effects of individual action are relatively unimportant. For similar reasons, we find that there is a limit to individual freedom in large economies as well. There is an urge in large economies to monitor or control those who can exert much power, such as monopolies, because they are important relative to the size of the economy. And in large economies some units remain relatively small and have partnership characteristics, such as the firm, which explains why authority and social control prevail.

Although tolerance is a virtue it should be noted that it is coupled with indifference. And even if the defence of individual rights is a noble fight the material preconditions should not be ignored. It is the absence of these preconditions that made it difficult for the mediaeval village and city to permit tolerance and unbounded individualism.

The argument that the extent of individual freedom is positively related to the size of the economy and inversely related to the power of the individual also reverses the causal order in the traditional explanations. The traditional accounts stress the emergence of a new institutional setting and/or a new mentality as factors that direct self-seeking individualism towards socially desirable ends and which subsequently create a dynamic upturn of an economy hitherto characterized by inertia. This line of reasoning however misses the point: in fact it is only at a sophisticated technological level that economies become large enough to be able to cope with and exploit the beneficial effects of individualism – that is, to make it socially progressive.

2.7 The Problem of Institutional Change

To show that the specific institutions discussed above have favourable consequences does not constitute a functional explanation. A proper functional explanation involves three distinct arguments (Halfpenny, 1981). In a functional formulation of historical materialism these are (1) a specification of the social institutions that are to be explained by (2) their favourable consequences (i.e. their function) for technological progress given the character of the technology, and (3) a propensity postulate, i.e. an assertion that the social system tends to develop technologically. Unless (3) is argued carefully the whole explanation easily becomes tautological. An elaboration of (3) should preferably specify mechanisms that make it plausible.

Functional explanations are most often and successfully used in evolutionary biology where the equivalent of (1) can be a feature in a species that has (2) some favourable effect for the reproductive capacity of the members of that species given the fact that (3) there is a natural selection process biased against species with unfavourable features. In this explanation there is reference neither to intentions nor to awareness of the functional relationship and the selection applies to the entity that has the feature rather than the feature as such. For these reasons the relevance of biological analogies has been considered small in the social sciences. Recently, however, there has been a growing interest in explanations of a similar type, for example explanations of profit-maximizing behaviour as the outcome of market selection: only entities (firms) that for some deliberate or non-deliberate reason perform profitably survive (Alchian, 1950; Nelson and Winter, 1982). When it comes to larger entities such as nations or cities or regions this type of explanation does not make sense, however.

A more promising functional variety should appeal to a reinforcing mechanism, a filtering or selection process that involves a human agency, in favour of features that are recognized but not necessarily anticipated as having some beneficial effects. This is usually called the welfare view (Nagel, 1977). According to this approach a feature or item, e.g. an institution, is explained by its welfare effects and it is assumed that there are mechanisms which direct society to states characterized by higher welfare. This type of explanation has some interesting properties that suit the arguments pursued in this book. Generally speaking, technological progress has the effect of increasing the welfare of the members of a society either in the form of higher consumption and/or more leisure (play, art, ceremonies), and an increased capacity to cope with the constraints of nature. As a rule we take it that man prefers more welfare in this sense to less. There may be an evolutionary element in the emergence of new institutions favourable for technological progress in the sense that they are not intended or anticipated but occur by chance and are selected once their beneficial effects have been vaguely recognized. Institutional innovations may thus have a similarity with the process of technological progress as it has been pictured here, i.e. as the outcome of stochastic processes. The criteria of selection of institutions and their effects on technological progress will be unrelated to the process that generates institutional 'mutations' (see van Parijs, 1981, especially ch. 2); chance variation and (semi-) deliberate selection, in other words.

A case in point can be the discussion above which focused on the fact that the emergence of scattered fields in open-field agriculture may very well be partible inheritance, while the explanation of its persistence may be a functional one. Once existing, the favourable effects of the open-field system – given the character of the prevailing technology – were recognized and households reinforced that structure of land-holding.

There are, however, several intricate problems related to the use of functional explanations. How can, in the first place, the favourable consequences be recognized if technological progress is very slow, almost unrecognizably small? The frequent reference in sloppy functionalism to long-term consequences is illegitimate. But there is another solution at least so long as we deal with the elementary social conditions for technological change in that these also produce immediate and recognizable side-effects. The elementary social conditions were presented as those contributing to cultural continuity and social order which – because of these effects – would also induce (slow) changes in technology through a process of accumulation of knowledge. Since social order and cultural continuity are desirable in their own right a selection of technologically progressive institutions may be reinforced. Returning again to open-field agriculture, the monitoring of production typical of the open-field system had both the short-term effect of increasing the productive efforts to a socially desirable level and, as a consequence of this, an increase in the capacity of the economy to cope with risks, which in turn contributed to cultural continuity.

Another important objection to functional explanations of the welfare variety is that it is not self-evident that technological progress brings benefits to all members of societies. As long as we consider fairly unsophisticated economies this may be an irrelevant objection because class divisions do not interfere with the distribution of the gains from technological progress. Similarly, when the resistance to institutional change originates from the dispossessed, as in the case of the enclosures, we would expect them to have little influence against the interests of the land-holders and large sections of the peasantry; political and economic resources are intertwined.

With respect to the mediaeval European economy, however, the objection cannot be dismissed that easily. How can we know that a powerful minority does not block institutional changes that would stimulate technological progress if the institutional changes erode their position? In particular, the lords monopolized the means of coercion and we would expect them to have resisted institutional changes unless they got a share of the increased production generated by technological progress. In chapter 3 there is an exposition of the defeudalization process along these lines. It points out that centuries of peasant struggles for independence and rights in land and their own labour, which is generally believed to be favourable to technological progress in agriculture, yielded results when population pressures on land increased land values to such an extent that lords found it in their own interest to succumb to peasant demands. Peasants gained freedom but at a high price, as a French historian aptly described the process.

A final objection concerns the stability of new institutions with favourable consequences. Since to a certain extent institutions have a public-good character, there are free-rider problems. The stability of an institution with favourable effects is therefore not necessarily self-reinforcing and it may in fact be in the interest of individuals to break an institutional rule while others comply.

Individualism is a case in point. If others show restraint or altruism it may be rewarding – in the short run – for an individual to act aggressively even though all would be worse off if all acted aggressively. Although the incidence of feud and conflict and institutional failure cannot be denied, small societies do have one advantage in dealing with either deviant behaviour or lack of information. For once their small size is an advantage: in small socieities it is fairly easy to observe and control the behaviour of others. As argued in sections 2.4 and 2.5, open-field agriculture and the guilds served to enhance both the information and control.

In larger societies the need for control of individualism diminishes for reasons discussed in section 2.6. Contrary to the case of institutions that restrain individualism there is a self-reinforcement in individualism, i.e. if others act individualistically it is normally best for one to act in the same manner, even though sometimes the outcome is socially undesirable. Even if it was in the interests of a ruling class to supervise and control individualism, it became increasingly costly and difficult to do so when economies grew larger.

No conclusive arguments for a functional explanation of pre-industrial institutions have been advanced. The functional explanation seems, however, plausible. The standard criticisms against functional explanations as discussed above do not seem to be highly relevant for the type of institutions and societies that have been discussed in this chapter. The proposed explanation, nevertheless, still has a conjectural character.

NOTES

1 See also the survey by Furubotn and Pejovich (1972).
2 It is worth quoting Frank Hahn at some length to illuminate the point just made. Discussing the limitations of economic theory he argues:

> The theory has a lively sense of original sin – all people act entirely in their self-interest quite narrowly defined. But, if that is so, will not individuals or groups of individuals seek to find ways to exert market power? By market power I mean a situation in which an individual's action can influence equilibrium prices. How can we be sure that the hypothesis that individuals act as if prices were given is not in conflict with the postulate that they are rational self-seeking agents? The answer is that we can only be sure if there is no market power for individuals to exploit. This can be shown to entail the condition that everyone in the economy, other than a given agent, can do as well when the agent trades as when he does not; this must be so whoever the given agent is. In general, this 'no surplus' condition will only be satisfied in 'large' economies... When market power is present the Smithian vision of the invisible hand is lost. Instead of a machine-like response of agents to prices, the agents will find themselves engaged in a game. That is, it will be necessary for them to take account of the decisions of other agents and, in particular, they may have to consider how these decisions are affected by their own. Their choices will now be among strategies. Here, economists are not agreed even what the appropriate notion of an equilibrium should be. But it becomes easy to show that plausible equilibria are no longer Pareto-efficient.... If the 'no surplus' condition is not met there must be an externality, almost by definition; that means that externalities are implicit in any departure from perfect competition. This seems to imply that one cannot ascribe failures of the invisible hand in the face of externalities exclusively to defective property rights. (1982, pp. 6–8)

3 See Hahn (1982) for a non-technical survey of the recent advancements in economic theory on which the preceding paragraphs are based. Cf. also Schotter and Schwödiauer (1980, pp. 500–2) for a discussion of the deficiencies of the property-rights approach in dealing with externalities.
4 Service (1966) gives an excellent survey of such arrangements among hunter-gatherer cultures.
5 It should be pointed out that public goods complicate the sort of institutional setting imagined by the property-rights paradigm. One immanent threat to cultural continuity has already been discussed in relation to the implications of the property-

rights approach. When production involves exhaustible resources, which it normally does, the cultural continuity of a society could eventually reach a state of crisis due to over-exploitation of that particular resource unless some institutional arrangement was reached concerning the access to that resource. The adequate alternatives to cope with that situation all involve some type of exclusive right in that particular resource, be it private or collective. There is a dominant tradition within the property-rights school that would claim the superiority of private property rights but historical evidence suggest that collective property and some regulation of individual access to that resource have been prevalent and suitable. This controversy can, however, be put aside because what is at issue is that exclusive rights, private or collective, require conventions and mechanisms of protection. That is, some sort of social organization that controls and monitors the contributions to the production of that public good.

6 See, for example, Cashdan (1985) and Schneider (1979), but the general theme is laid out in the classical study on this subject, Marcel Mauss's *The Gift* (1954, e.g. pp. 1, 63).
7 The hypothesis advanced by Dahlman also stresses the importance of the growth of the market but puts undue stress on an alleged specialization of villages into either arable or livestock production. Since the subdivided fields are related to the arable–pasturage duality the system declines with the specialization endorsed by the growth of market. There is a chronological association of enclosures and the growth of the market but it was a lengthy process: after all the enclosures were for most countries not completed until (well into) the nineteenth century. Furthermore an orientation of households away from the village community and towards the market and a subsequent social differentiation is beyond doubt. But the clear-cut specialization between arable and livestock production at village level, suggested by Dahlman as a sufficient cause, is not corroborated by the English experience (Yelling, 1982) or by evidence from Scandinavia.

3 Growth and Stagnation in the European Mediaeval Economy

3.1 Introduction

The possibilities of, and constraints on, sustained growth in Europe during the mediaeval and early modern period occupy a large and growing literature. Within that literature, at least three different strands can be identified: the analysis tends to stress the importance of demographic factors, the effects of property relations and distributive struggle, or the role of trade and urban growth.

The demographic approach claims support from Ricardo and Malthus and is presently associated with economic historians such as Habakkuk (1958), Postan (e.g. 1972, 1973) and Le Roy Ladurie (1966, 1982, pp. 72–8). Proponents of the approach typically take a rather pessimistic view as regards technological change. The feudal system is seen as technologically stagnant and it is argued that the constraints on growth are largely determined by demographic forces: with increasing population and limited land the relative share of labour will fall, whereas decreasing population will lead to an increasing relative share of labour and a new wave of expansion.

The property-relations approach is clearly inspired by Marx and is now associated with Dobb (1946), Bois (1976) and Brenner (1976, 1982). According to this approach, the dynamic properties of pre-industrial economies will depend on the nature of the property relations and on the relative strength of the different classes. Although the authors referred to above express different opinions on some issues they all argue that the effects of demographic changes are always mediated through the social relations typical of the mode of production. The modernization process cannot proceed unless feudal social relations are replaced by capitalist farming (capitalist farming being a concept sufficiently vague to cover the spectrum from market-oriented independent farmers to large-scale agrarian entrepreneurs with hired labour). While Brenner views the capitalist agrarian entrepreneur with hired labour as an essential agent in agricultural modernization, others within the property-relations school more wisely tend to point out the dynamic capacity of a free and independent peasantry and a tenantry with well-defined rights.

The commercialization approach finally suggests that expanding trade and urban growth are associated with increased specialization and that this in turn induces more efficient economic organization in all sectors of the economy. Theoretically, this approach owes much to Adam Smith and his emphasis on the relation between the size of the market and the division of labour. Among modern exponents of the approach are Pirenne (1925, 1936) and Sweezy (1950), who both stress the inventive and liberating role of the market.

It is perhaps in the nature of academic discussions to stress the difference between alternative approaches or even the incompatibility of the approaches. This is so particularly when participants in a debate choose to associate themselves with the grand theorists of the past, and when strong ideological questions are involved. The so-called transition debate in the 1950s following Sweezy's critical review of Dobb's *Studies in the Development of Capitalism* (1946) is one example, and polarization of the views also characterizes the debate which followed Brenner's critical appraisal of agrarian historiography in *Past and Present* (1976).

The different approaches to mediaeval growth should not, however, be seen as mutually exclusive. On the contrary, they complement each other and each of them may help to highlight important aspects of the period. The purpose, therefore, of this chapter is to combine elements from all three approaches in a synthetic model and to analyse the preconditions for sustained growth as well as for crisis and stagnation within this more general framework.

The outline of the chapter is as follows. Section 3.2 starts by describing the main complementarities between the rival approaches; section 3.3 traces some of the consequences of adopting a synthetic approach on the notion of feudalism; a verbal description of the model is given in section 3.4; the implications of the model are discussed in relation to empirical data in sections 3.5–3.10; and finally, there is an appendix with a formal version of the model and its implications.

3.2 Common Ground and Interrelations

It should be noted that all three approaches are in agreement on some issues. It is agreed, for instance, that population growth is Malthusian in the sense that it is related to per-capita income. It is also accepted by all parties that there is a limited supply of (good) cultivable land. This however does not necessarily set the stage for a simple Ricardian analysis in the demographic tradition. First of all, it is not obvious that the most fertile land is brought into use first. Very fertile land which is situated far from existing urban areas may remain unused (or even unknown) until population pressure causes migration and the formation of new population centres. The very notion of a simple one-dimensional ranking of land in terms of fertility is also extremely suspect when the existence of several crops is taken into account.[1] The increased specialization, stimulating new crops and rotation systems, stressed by the commercialization approach, will make this argument very important.

At the centre of the controversy are the effects of property relations and population growth on the evolution of the mediaeval economy. In the property-relations approach, the key to economic development lies in changes in the feudal mode of production which shift the distribution of income in favour of peasants. But even if property relations change and thereby affect the distribution of income, this does not in itself explain growth or stagnation. It is the interplay between the distribution of income and the incentives and resources for technological change which is of interest in a dynamic theory. According to the property-relations approach, the main constraint on technological change is the availability of resources for investment and consumption. I will argue that the main effects of changes in property relations lie elsewhere than in the distribution of income. More important are the effects on the incentives and commitment of producers, on their freedom of action and ability to respond to market opportunities. These incentive and commitment effects are, furthermore, independent of the change in distribution. A change in property relations may, as we shall see, simultaneously give peasants greater freedom and shift the distribution of income in favour of the land-holding classes. Incentive effects are thus immune to criticisms which question the effects on distributive shares of the loosening of feudal bonds (cf. section 3.9).

Incentive effects are related to (but not identical to) some elements in the commercialization approach. The commercialization approach argues that apart from the informational and ideological aspects of trade and commerce, there are important 'technical' effects of increased specialization. Increased trade in agrarian products will have a land-saving bias in agriculture because it enables different regions to specialize in the products which are best suited to their particular advantages in terms of soil and climate, access to markets and transport. Exchange between urban and agrarian producers also involves both static and dynamic specialization effects: productivity increases when a producer concentrates on fewer products, and the repetition of specialized tasks is also likely to induce faster learning by doing, or 'practice makes perfect' as Marshall put it. The process just described is essential in the commercialization account of the modernization process but it presupposes profound changes in property relations. Commercialization and the associated increase in specialization and trade would be impossible unless peasants have gained relative freedom of action and choice. Changes in property rights and commercialization thus interact and the two approaches are complementary rather than mutually exclusive.

There is also a close interdependence between changes in property relations and population growth and this interdependence makes it futile to present the demographic and the property-relations approaches as mutually exclusive. In an economy with free land and free men, and where land is abundant in relation to labour, the marginal and average productivities of land are identical. No landowner could attract free men to till the land unless he paid them their entire product because that is what they would produce on the freely available land. Even if all available land was partitioned among a class of lords they would compete for the scarce labour so as to drive agricultural wages to the level of the average productivity; it would be impossible to extract rent from peasants.[2]

In post-Roman Europe there was a general shortage of labour relative to land. Enserfment and the direct restriction on the freedom and mobility of peasants was thus a precondition for the extraction of rent. The right of lords to income from fiefs (as payment for services to fragile central state powers) would have been worthless without enserfment. The basic function of serfdom was to prevent peasants from taking free land into possession and to restrict inter-lord competition for labour (see Domar, 1970 for an analysis of serfdom along these lines).

Whatever contributions this system made to Europe in terms of order, social stability and, indirectly, population growth and the start of a self-sustaining growth process, it had drawbacks as a system of agricultural production management.[3] Monitoring of labour was costly and work motivation and efficiency were low. The system nonetheless prevailed as long as land was in abundant supply. Continuing population growth, however, changed the man/land ratio to the advantage of lords. The growing competition for land gradually made lords aware of the fact that they were in possession of marketable assets. In consequence, the cash nexus replaced personal obligations and the manor disappeared from the most advanced areas of agrarian transition. Changes in property relations were thus intimately linked to population pressure; population pressure and the shortage of land induced endogenous changes in property relations. Centuries of peasant struggle gave results when lords found it in their own interest to succumb to peasant demands. Land became private property and owners of land claimed a price and rent depending on the market configuration rather than political authority (although the latter remained vital for the defence of property relations as such). A 'business-like relationship', as an historian of the agricultural transformation in the Low Countries prefers to call it, replaced the relation of direct subordination (van der Woude, 1982a, p. 195).

3.3 What and When was Feudalism?

In this process of transition the old nobility sometimes lost and sometimes gained property. But even when they saw their fiefs dwindle that did not always mean that land was transferred to peasants. Urban proprietors showed an increasing interest in land when population made it a scarce commodity. This new relationship between peasants and urban and rural landowners is not feudalism. For sentimental, political and theoretical reasons the concept of feudalism has always invited controversy and there is little agreement on the precise meaning of the word. According to the view defended here, it would be theoretically sound to restrict the use of the concept to the politico-legal system in which peasants had personal obligations in terms of labour dues to lords, the nature of these obligations amounting to serfdom, and in which lords held land as fiefs rather than personal property.

But if the emergent system is not feudalism then what is it? Capitalism? In common usage, capitalism implies private property and markets for capital, goods, land and labour. The system which we have just described and which was fully developed in important areas of Europe by the end of the thirteenth century, had markets for land, capital and goods and to a certain extent labour. But the loosening of personal obligations to lords strengthened the household as the basic unit of production sometimes with one or several servants. Insofar as one is hesitant to call a market-oriented system based on household production in urban industry capitalist, it would be wise not to widen the use of capitalism to its agrarian counterpart. The concept of capitalism will consequently be reserved for production for markets by large-scale production units, i.e. firms. The Marxian term, petty-commodity production, is perhaps the best term for the system we are considering; it accentuates the small production units but at the same time stresses the fact that it is production of commodities for markets by means of inputs obtained through markets. The mediaeval economy, in our view, experiences a transition from a feudal mode of production to petty-commodity production.

3.4 Property Relations and Technology in Agrarian Transition

The model is described formally in an appendix to this chapter, the purpose of this section being to provide a verbal description of the main features and implications. Under consideration is a predominantly agrarian economy with a small urban sector and with labour mobility between sectors. The basic unit of production is the peasant household paying rents in money, in kind or in labour (performed on the demesne) to landlords or urban proprietors. The peasant household produces agrarian goods using capital and labour, the former originating in the urban sector. Labour input is measured in efficiency units and it is assumed that the efficiency of labour depends on per-capita consumption of agrarian goods. This specification of labour input captures the problem (highlighted in the literature on development economics) that a clear-cut division between production input and consumption cannot be made. With increasing consumption of food, workers become more efficient; they are stronger and able to work harder and more intensively.

Peasant net income, i.e. income net of rents (in whatever form they are paid), is spent on urban and agrarian goods. It will be suggested here that as net income increases peasant households will devote an increasing share of income to urban goods. It is assumed, furthermore, that capital goods form a constant fraction of urban goods purchased by peasants. Capital investment will thus increase as net income increases.

Two additional factors are of the utmost importance in peasant production. There may be diminishing returns due to the fact that land is in limited supply but against this must be set technological change. The relative strength of these two forces are, as we shall see, critical for the outcome. The diminishing returns argument is straightforward (although, as mentioned above, the increased specialization associated with market involvement may weaken the tendency of diminishing returns). Technological change, however, is not an exogenous factor and a few comments may be needed.

In chapter 1 it was argued that production endogenously creates technological change; small technological mutations occur randomly and some deliberate experimentation also takes place. The diffusion of technological knowledge will depend on the intensity of contact between the individual peasant household and the outside world. Servants played an important role in the transmission of knowledge because they were very mobile, learning a specific skill in each employment.[4] It is plausible that the intensity of contact between households is related to the degree of market involvement and that the technological inertia seen in self-sufficient agriculture is due primarily to lack of information and exposure to new knowledge rather than to lack of interest. Societies with low per-capita income experience slow technological change because households have little to trade, or lack rights to do so, and remain isolated. Pre-industrial economies improve methods by selecting best practices from random mutations of established methods and through trial and error. If the number of producing units who exchange information increases, which will happen when trade brings people together in larger networks, then the number of such beneficial changes known to the community will also increase. The speed of technological change is thus endogenous. Rising net incomes not only increase capital investment by the peasant household, they also lead to stronger and more frequent contact with the outside world through trade. This enhances the diffusion of information which in turn affects the rate of technological progress as well as the ratio of average to best-practice productivity. Increased trade associated with a rise in incomes thus causes a shift in the technological-change parameter. Furthermore trade and population growth both contribute to making the economy larger – i.e. increasing the number of individuals or households participating in the economy. As was pointed out in chapter 2, the size of an economy is crucial for the social acceptance of a more individualistic outlook which under certain circumstances will contribute to economic progress.

Turning now to urban production, it is assumed that there are increasing returns. A higher level of demand and production allows for an increased division of labour (cf. the detailed discussion of the Adam Smith effect in chapter 1). As in agriculture, there is technological progress due to learning by doing, random mutation and purposeful experimentation. Urban prices are flexible relative to agricultural prices and there is an inverse relation between prices and productivities. If, for example, productivity in the agrarian sector grows slower than in the urban sector then there will be an increase in relative agrarian prices. The rationale for the link between productivities and prices is labour migration. With the gradual loosening of the feudal bonds the mediaeval economy experienced considerable mobility between sectors and sustained differences in income levels induced migration. The sectoral shifts may not have been very fast but in a model of secular trends it is appropriate to assume the equalization of rural and urban income levels.

3.5 Implications

The assumptions described above are capable of generating a variety of growth paths. Assume, for instance, that there is initially unexploited land. Even in the absence of technological progress this would lead to increasing per-capita income. Population growth at constant labour productivity will increase the aggregate demand for urban goods (i.e. non-agrarian goods). With increasing demand for urban goods the ensuing process of specialization will increase labour productivity in urban production and lower the prices of urban goods in terms of agrarian goods. Real per-capita investment and hence income in agriculture is raised and this leads to increased per-capita income in agriculture. The proportion of income spent on urban goods increases with income and that implies that both the amount of capital goods per agrarian household and labour productivity will increase further. An initial push in terms of population growth in the agrarian sector will therefore lead to income growth in both sectors. Increasing income will be associated with an ongoing process of specialization and market involvement. Since technological change is related to the intensity of producers' contact with the outside world this may in turn induce faster technical progress; there may be a shift in the parameter describing the rate of technological change.

Not surprisingly the economy portrayed above will exhibit perpetual growth of population and real per-capita income if diminishing returns are weak in relation to technological progress and increasing returns in urban production. The urbanization ratio will increase. But the expansion process may encounter problems and the increase in per-capita incomes may come to a halt or even be reversed. As population continues to grow – presumably at a higher rate because of the increase in per-capita income – the problems of a limited supply of land may reassert themselves and the outcome depends on the relative strength of diminishing returns in agriculture *vis-à-vis* the factors making for continued growth, i.e. increasing returns in urban production and technological change in both sectors. *A priori*, not much can be said about the quantitative strength of these various factors and there is therefore no reason to expect one unique development path from mediaeval to early modern agriculture.

If diminishing returns are or become strong in relation to the countervailing forces the economy will settle in an equilibrium state of positive population growth with stagnant per-capita income. The urbanization ratio will be stable as well. In the case where technological progress is entirely ruled out the long-run Malthusian equilibrium would be characterized by stagnant population and constant per-capita income. This is also the type of equilibrium imagined by proponents of the demographic approach such as Postan and Le Roy Ladurie. It must be pointed out, however, that there are many Malthusian equilibria each characterized by a specific configuration of per-capita income and population growth ultimately dependent on the strength of the technological-change parameter at given decreasing returns. The demographic approach in the Postan–Le Roy Ladurie version, in which technological change is insignificant, is in fact but a special case of a more general Malthusian model. If technological change is allowed for the equilibrium, per-capita income is above subsistence and permits population growth.

As has been pointed out, there are endogenous forces which may cause an upward shift in the rate of technological change. Feudal social relations are dissolved under the population-pressure-induced increases in land values. The intensified market involvement of producers enhances effort and the diffusion of knowledge. Thus the model suggests the possibility of an increase in per-capita income over time. That is so because an upward shift in the technological-change parameter – *ceteris paribus* – will move the economy from one equilibrium income (and its associated population growth) to a new and higher one. In this regime, the positive aspects of population growth, the stimulus to demand and hence increasing returns in urban production and the incentive and efficiency effects of the loosening of the feudal bonds generated by the increased man/land ratio, dominate the negative aspects, diminishing returns. Rising real incomes imply that expenditure on urban goods increases as a proportion of total incomes and since per-capita incomes are equal in the two sectors this then implies a relative increase in the urban population. The rising degree of urbanization is thus a characteristic feature of this regime.

Although this model of a growth process has some Malthusian (and Ricardian) characteristics – i.e. diminishing returns and Malthusian population growth – it is genuinely un-Malthusian in its argument that population growth endogenously creates countervailing forces to diminishing returns. The approach defended here can exhibit an increase in per-capita income explained as an upward movement of a steady-growth equilibrium – i.e. as a movement from one equilibrium income to higher ones. We will call such a trajectory a self-sustaining growth process.

A change in the distribution of income in favour of peasants will not divert the economy from its long-term equilibrium. It will only provoke a temporary increase in income and population growth but eventually the economy returns to its original per-capita income and the associated lower rate of population growth. The standard criticism of the demographic approach has been its alleged neglect of social relations and their impact on the distribution of income. Given the assumptions of Postan and Le Roy Ladurie about dominating and strong diminishing returns and insignificant technological change, their position is tenable. The position taken by R. Brenner seems therefore misconceived insofar as it localizes the detrimental effect of feudal agriculture in its distributional consequences. In a long-run Malthusian equilibrium the real per-capita income of peasants is independent of the distribution of income. But if the changes in distribution, which may be unfavourable to the peasants, are associated with changed social relations which cause a shift in the rate of technological change, then the per-capita income may diverge positively and permanently from its former equilibrium level.

If decreasing returns are small in relation to technological change and increasing returns, then the most intuitively plausible growth path is one of permanent increase in per-capita income, and there is no stable equilibrium. We have already dwelt upon this implication when discussing the early phase of the European mediaeval expansion. With near-constant returns to scale, which might apply after the Black Death and subsequent epidemics increased the land/man ratio, there is an equilibrium, but it may be unstable. Again the most intuitively plausible expansion path would be one of sustained growth. A simple demographic model therefore suggests a new wave of expansion of population. This does not occur. In section 3.10 this paradox is discussed and, to some extent at least, explained within the general framework defended in this chapter.

It should be clear by now that the evolution of the mediaeval economy will be related to the extent of the diminishing returns, to what extent such tendencies were countervailed by increased specialization and the capacity of the system to generate technological change. *A priori*, not much can be said about the relative strengths of these forces but they must be assessed in the light of historical evidence. The implications are discussed in that context.

3.6 Urbanization and Population Growth in Europe

Europe experienced many troubled centuries after the decline of the Roman Empire. Population declined, trade routes were disrupted and many towns were reduced to insignificance. Central authority broke down and was at best replaced by local order; but often that failed as well. By the tenth century, or perhaps earlier, things began to change. Political and social conditions became more ordered and population

TABLE 3.1 Population in Europe (millions)

	Years				
	1000	1300	1400	1500	1600
Western and Central Europe	12	35.5 (1340)	22.5 (1450)	36.2	50.5
Italy	5	10 (1340)	7.5 (1450)	10.5	13.3
France and Low Countries	6	19 (1340)	12 (1450)	18.3	21.3
of which Low Countries	–	–	–	1.9	2.9
of which Netherlands	–	–	–	0.9–1	1.4–1.6
British Isles	2	5 (1340)	3 (1450)	4.4	6.8
of which England	2.5–3 (1086)	5–6 (1347) 2.5–3 (1370)	2–2.5 (1450)	2.25–2.75 (1525) 3 (1551)	4.1 (1601)

Comment: Needless to say the figures are approximations. Time series are based on different sources and the comparability over time is reduced.

Sources: Western and Central Europe, total and national figures: Russel (1972, p. 36), and Mols (1974, p. 38). Netherlands: Slicher van Bath (1972, p. 337). England: Hatcher (1977, p. 69), Wrigley and Schofield (1981, pp. 208–9), and Hatcher and Miller (1978, p. 28).

increased again. By the eleventh century we witness the beginning of a new dynamic era in European history. The following three centuries show rapid population growth interrupted only by occasional famines. In some 300 years the population of Western and Central Europe tripled, as can be seen from table 3.1.

The figures are, needless to say, approximations so it is unwise to dwell on the regional differences in rates of poplation growth. The relative size of populations of the different areas at the beginning of the fourteenth century is, however, believed to be fairly accurate. France is a very populous country by the beginning of the fourteenth century and very densely populated too with some 35 inhabitants per square kilometre. This, however, was still only half of the density found in Flanders and northern Italy.

The most important indirect evidence of a rise in income apart from population growth is the increase in the urbanization ratio – i.e. the ratio of urban to total population. The urbanization ratio is an approximate measure of the relative size of the non-agrarian sector and hence, recalling the logic of our model, an indicator of labour productivity in the agrarian sector: an increase in the urbanization ratio can be interpreted as an increase in labour productivity and, under plausible assumptions, as an increase in per-capita income of peasants. The logic of the explanation should perhaps be repeated at this stage. In a closed economy the agrarian population produces food for the entire population. An increase in the urban proportion implies that an increased proportion of aggregate production is spent on urban goods (i.e. all non-agrarian goods and services). The growth in demand for urban goods is caused by an increase in per-capita income: with an increase in per-capita income a declining proportion is spent on necessities (i.e. food) and a larger fraction on urban goods. In order to get a detailed estimate of the productivity gain associated with a given increase in the urbanization ratio one needs to specify the marginal propensity to consume agrarian goods and, if average propensities differ between social classes, the distribution of income. This will be clear from the formal description of the model and from the procedure followed for the productivity estimates made in chapter 4.

We still have the visible and technologically sophisticated effects of the wave of urbanization that characterized the years from the eleventh to the fourteenth century. Most existing cities of some importance were either founded (the majority of them) or rescued from oblivion in this period. Table 3.2 shows collected data that refer to the size of the urban population as a percentage of total population. There are important

TABLE 3.2 *Percentage of urban population to total population (urbanization ratio)*

	Years		
	1000	1300	1500
Western and Central Europe	$\leqslant 5$	10–15	10–15
Low Countries	–	25–30	45
France	10 (1200)	15	–
England	–	15–20	15–20

Comment: These figures are based throughout on a generous but from our point of view congenial definition of urban population as those inhabiting agglomerations, whatever their size, which were basically engaged in non-agrarian production. If urban population was restricted to cities of a substantial size, say above the size of 3000, the urban population ratio would probably be half as big (cf. Bairoch, 1985, chs 8 and 10).

Sources: For the Low Countries see van der Woude (1982, pp. 55–72). England: Clay (1984, p. 165), see also Goose (1986, pp. 165–85). France: Sivéry (1984, pp. 13–14). Europe: Thrupp (1972, p. 263).

differences. The Low Countries in particular have an extremely high urbanization ratio. One possible explanation could be that this region satisfied a large proportion of its agricultural needs through imports. The evidence suggests, however, that this area, including northern France, was largely self-sufficient in food at least up to the fourteenth century. It was not until the fifteenth and sixteenth centuries that long-distance trade supplied a significant part, say more than ten per cent, of the corn needs of the area. At that time the urban economy had reached an even higher level of sophistication and urban population constituted almost half of the population. Nevertheless, the urbanization was associated with an intensification of interregional and international trade as well as diversification of urban manufacturing for urban and agrarian markets and we return to this issue below.

The urbanization ratio stagnated with (or before) the outbreak of the Black Death and in some regions stagnation had probably already set in by the end of the thirteenth century. We know very little about what happened between 1350 and the beginning of the sixteenth century but it is commonly assumed that the urbanization ratio was more or less unchanged. Possible exceptions are the more dynamic regions of Europe such as the Low Countries and northern Italy, where urban life recovered more swiftly from the devastating effects of the Black Death. Then in the sixteenth century came a new wave of urbanization, though less marked in France than in other parts of Europe.

3.7 Malthusianism Revisited

There is general agreement that the early phase of the mediaeval expansion, as well as the second phase starting at the end of the fifteenth century, have the characteristics which in our model make for permanent growth. Land is in abundant supply, population increases and so does the degree of urbanization. When population pressure mounted in the twelfth and thirteenth centuries we may expect diminishing returns but also population-pressure-induced institutional changes that stimulated technological progress, substitution of labour and other inputs for land and specialization. Within the urban population there was increasing specialization, a division of labour which manifested itself in a rapid growth in the number of occupations and trades. This process continued well into the thirteenth century in most parts of

Europe and it may be characterized as a self-sustaining growth process – i.e. as a process of growing per-capita income, urbanization and population. The underlying changes in terms of technological progress in agricultural implements and methods included changes in the exploitation of traction power by means of better harnessing, more widespread use of horses with a nailed horseshoe, irrigation and control of soil humidity through ditches and dikes, better ploughs, extensive use of iron in tools, new rotation systems, new crops, and manuring that led to a gradual suppression of the fallow – cf. van Houtte (1980) and chapter 1 for a detailed discussion.

There was also, and it is so well known that we do not have to dwell upon it here, a gradual regeneration of intellectual life: learned institutions appeared, interest in natural and empirical sciences flourished and occupational specialization developed. Accompanying the increased trade there was also an evolution of institutions such as credit and insurance. The claim of an accompanying intellectual and institutional evolution – although essential for the general argument pursued in this chapter – will not be substantiated here because it is fairly uncontroversial and since there are recent surveys of this very topic (cf., for example, Wolff, 1986).

There are, however, two interrelated problems that must be dealt with. These concern the interpretation of the period before the outbreak of the Black Death in 1348, and the prolonged demographic decline in the century after the first epidemic wave. While few would dispute that Europe experienced a period of unprecedented growth of population and income in the eleventh, twelfth and thirteenth centuries, there is little consensus as to the true character of the hundred years before the outbreak of the Black Death.

The demographic approach tends to view this period as the end of the expansion. The pressure on land increased, available land was of inferior quality and good land deteriorated because of over-intensive cultivation and insufficient manuring. That the soil was becoming exhausted is an important element in Postan's hypothesis and it is argued that the hypothesis was later substantiated by Titow's (1972) study of the Winchester estates. A closer and more sophisticated statistical analysis does not reveal such a trend, however (Desai, 1977). In short, accepting this approach, Europe headed towards a crisis of subsistence which manifested itself in the increased frequency of famines and finally in the severe effects of the Black Death, although the actual outbreak of the disease is normally seen as unrelated to the alleged distress of the period.

It will be argued here that the demographic interpretation cannot be upheld as a general model for European history, and its validity for English economic history can also be challenged. The basic presumption of the demographic approach is that technological progress was not strong enough to prevent the diminishing returns to cause a fall in incomes and population. There is now, however, a growing body of literature that stresses the advanced nature of agricultural institutions and methods in several distinct European regions emerging from the second half of the thirteenth century and continuing into the fourteenth century.

Schematically, the changes in methods can be described as (1) substitution of labour, implements (capital), and manuring for land, (2) technological changes with a land-saving bias such as the suppression of the fallow, intensified land use by means of irrigation and improved quality of land by means of leguminous crops (adding nitrogen), improved rotation schemes, better strains, new plants and an improved match between crops, soils and climate induced by trade and regional specialization, and (3) technological changes with a labour-saving bias embodied in new implements, such as spades and ploughs. These recent findings represent a challenge to the demographic approach because it is now clear that important regions developed differently from the distress supposedly characterizing the whole of Europe.

3.8 Regional Diversity

There was in fact early opposition from continental economic historians such as Carlo Cipolla who argued against efforts to make the demographic model fit Italy (Miani, 1964), and Abel (1980a, pp. 40–3) raises doubts to such an interpretation as well in his classical study of European agriculture. Italy was not, of course, unaffected by the general crisis in the European economy which followed the demographic disasters in the middle of the fourteenth century but early fourteenth-century northern Italy did not exhibit any signs of an impending crisis of subsistence. Some historians go so far as to argue that several regions in Italy underwent an agrarian revolution in the thirteenth and fourteenth centuries (e.g. Dowd, 1961, pp. 143–60). This optimism is not shared by all (see Rotelli, 1973). There were, however, spectacular changes – e.g. the reappearance of irrigation. In areas with sufficient water supply, ingenious systems of irrigation and canals were built and as a result meadows could yield grass up to seven times a year. The basis for a dairy industry was therefore at hand.

The economic progress shown by Italy was not unique. Recent research shows the existence of several progressive regions which were large enough to engage in a developed specialization of agrarian and urban production. Apart from northern Italy, the most prominent examples are northern France and the Low Countries as well as the Rheinstrasse, the region surrounding the lower parts of the Rhine and its tributaries (Derville, 1978; Irsigler, 1982; Sivéry, 1976, 1977, 1982; Verhulst, 1985). The crisis of the second decade of the fourteenth century was very much less severe in these areas compared to the rest of Europe.

G. Sivéry, discussing northern France, emphatically disputes the idea that the late thirteenth and early fourteenth centuries were just a prelude to the coming demographic catastrophe. Instead, he puts forward the thesis of a continuity of agrarian development in spite of the Black Death (Sivéry, 1982, pp. 667–81). The agrarian transformation during this period involved the cultivation of new crops for industry and for the growing consumer market in the cities. Gardening became an important element of peasant activity in the vicinity of cities. In gardening cultures flexibility *vis-à-vis* market opportunities were important. Open-field agriculture may have a disadvantage in this respect but gardening typically developed apart from the open-field structure. In the gradual suppression of the fallow, the cultivation of fodder crops and industrial plants for breweries and the textile trades were preeminent.

A similar thesis is advanced by other students of the agrarian history of the Low Countries and northern France, most notably H. van der Wee (Aerts and van der Wee, 1982, pp. 29–31; van der Wee, 1978, pp. 1–9). Not only is there ample evidence of the vigour of these economies well up to the Black Death, although none of them could have been unaffected by the severe climatological conditions in the second decade of the fourteenth century, but it also seems as if they were able to recover more easily and rapidly after the Black Death. In Italy there were, according to Cipolla, signs of a recovery with increasing investment in agriculture from the beginning of the fifteenth century (Cipolla, 1949, pp. 181–4).

In England it is believed that population growth continued until the famines in 1315–17 but it is not easy to establish a uniform trend after that date. There is evidence of decline in some parts (e.g. Essex, see Poos, 1985) but others have argued that neither the stagnating population nor the familiar signs of an agrarian crisis could be corroborated (Campbell, 1984, pp. 95–101; Harvey, 1966, pp. 23–42; Razi, 1980, pp. 94–8). Assuming Malthusian population growth a stagnation in population must have been preceded by a decline in income and, hence, the urbanization ratio. There are no claims as to a declining urbanization ratio (although it may have been stagnating) in early fourteenth-century England, however. Recent reappraisals of the size of major cities in the first half of the fourteenth century has led to a doubling of the size of London (to some 90,000) and some provincial cities compared to previous estimates (Keene, 1985). The continued proliferation of markets – serving mainly local trade – in the fourteenth century (although the rate of growth was lower than in the thirteenth century) can be interpreted as evidence of economic vitality (Britnell, 1981).

On the basis of the above it would seem rash to make unambiguous statements about the pattern of population growth in Europe in the second quarter of the fourteenth century.[5] We then have to rely on other indicators of economic well-being and distress. By and large we should be inclined to a positive assessment of the state of an economy which shows signs of innovation in methods, crops and institutions. Furthermore, it seems reasonable to assume that an increased diversification in agrarian production, such as the increased importance of industrial plants, meat, legumes and fruits, must respond to a sophistication of demand which seldom occurs in periods of declining per-capita income. This is, in fact, what actually happened in several regions in continental Europe in the hundred years before the Black Death, which has already been referred to in the preceding paragraphs. There are also some recent studies that emphasize the regional diversity of fourteenth-century English economic history. Eastern and perhaps south-eastern England developed an urban–agrarian complex very similar to the ones found in the progressive regions of continental Europe (Campbell, 1983a, b; see also Hallam, 1981, ch. 2).

Campbell goes as far as arguing that the two important components of the agrarian revolution of the seventeenth century in England – i.e. a fully developed system of convertible husbandry and ingenious methods of maintaining the fertility of the soil – were already present in parts of eastern England before the outbreak of the Black Death. As a consequence of the new findings we have to revise the dating of important agricultural innovations on the continent as well. Those who previously believed that the new methods (intensive use of land, livestock farming and a preservation of soil fertility through intensive input of labour, marling and manure) appeared much later now admit that the transformation was under way by the end of the thirteenth century and continued in the fourteenth century – i.e. during the very period when the dynamic phase of the European economy supposedly came to an end (see Verhulst, 1985 who revises the traditional view of Verhulst, 1963).

The advanced and dynamic regions share other important characteristics. Contrary to what one would expect from a demographic approach, they had a higher population density than most other regions in Europe by the beginning of the fourteenth century. Furthermore, the urbanization ratio was considerably higher than the European average: it reached figures as high as 30 per cent in the Low Countries and northern Italy. Land was used more intensively than in other parts but the yields were still at par if not higher than in less populous areas. In northern France and the Low Countries, yields per unit of land were twice as high as those believed to be typical on the other side of the channel (Derville, 1987; Morineau, 1977). Manuring, marling and rotation schemes including the sowing of pulses and fodder crops on an extensive scale contributed to cattle breeding and to the preservation of soil fertility. The vicinity of cities also stimulated an intensive gardening culture including vegetables, fruits and industrial plants such as flax, hemp and hops.

The vitality of these regions underlines the validity of the theoretical objections to the Malthus–Ricardian view spelled out above. Being basically a one-good model it pre-supposes a one-dimensional ranking of land in terms of fertility. In view of the developed specialization – i.e. an effort to exploit the specific capacity of different types of soil – this presumption breaks down.

Having, furthermore, a higher population density than other parts of Europe made it possible to increase the input of labour in production. It has been shown, for example, that the high yields in some areas of the Low Countries cannot be ascribed to the quality of the land but very often to the high labour input. This implies a lower productivity per hour of labour compared to a situation where land was more fertile or more abundant. But favourable demand conditions as well as advanced crop rotation systems requiring manuring and intensive working of the soil stimulated labour effort in the highly commercialized regions. This led to high levels of production per man-year and high per-capita income.

3.9 Incentives and Agrarian Transition

There are other characteristics of the technologically advanced regions that must be mentioned. Social relations gave peasants far more options for market involvement than in less-developed areas. Campbell speaks of the Norfolk peasant as more free than in other parts of England and supports the view that weak feudal bonds stimulated technological progress. There was in most of Europe a trend of demanorialization – i.e. diminishing importance of the manor as a production unit and a commutation of labour services to rents in kind or in money – accompanying the economic expansion. In France, this process was more or less uninterrupted and completed before the Black Death. In England it is believed to have been temporarily halted in the thirteenth century only to be continued again (Dyer, 1980, pp. 97–101).

Not only did farmers gain legal rights as regards the arbitrary actions of owners; institutions developed which fostered a mutual interest in improvements of equipment and methods (van der Woude, 1982b; Chittolini and Coppola, 1982). Urban credit penetrated the surrounding countryside establishing a market for land (Desportes, 1977, pp. 655–6; Fourquin, 1972, p. 138; King, 1973). Sometimes urban interest in agrarian activities was directly linked to the cultivation of industrial plants (Irsigler, 1983). Urban proprietors subletting land they had gained control over often originated from the countryside and had intimate knowledge of agricultural practice, which may explain why the non-efficient characteristics of absentee ownership were rarely observed.

As manorial jurisdiction over peasants weakened in the technologically advanced regions, peasant households, small and large, owners and tenants, became increasingly oriented towards the market with their produce and, to some extent, their labour. The relative freedom gained by peasants did not, however, imply that they became free of economic burdens. What happened was that dependence on a lord was transformed into dependence on the market. In fact, the price of freedom was high which probably explains why lords were willing to accept the commutation of labour services and granting of hereditary rights and sometimes full property rights to peasants. Simultaneously, the role of family and kinship in the transmission of landed property lost part of its importance and was supplemented by market acquisition (Smith, 1984, chs 1, 2 and 5).

The trend in demanorialization is consistent with the Domar hypothesis, which predicts a loosening of feudal bonds as the man/land ratio increases. But the geography of the defeudalization process is far too diverse to be encapsulated in an abstract hypothesis in which political and social inertia and customs are not explicitly acknowledged. It still has strong heuristic power, though. It was not the increased markets for agricultural goods or the monetarization of the economy *per se* that dissolved the manor as the paramount production unit. The high population density, the ensuing land scarcity and, as a consequence, high land values made the market forces work to the advantage of the land-holding class. As long as land was in good supply the feudal bond presented the only way in which lords could get an income from their land. Manorial production was then in the interest of lords even though the manor was often an inefficient production unit due to high supervision costs and low motivation among the labourers. This probably explains why feudalism persisted or reappeared during the so-called second serfdom in those regions, in particular Eastern Europe, where land values and population density were lower (Domar, 1970). Predictably, there was a seigneurial reaction in Western Europe after the Black Death and the drastically reduced man/land ratio. It did not succeed, partially because peasant freedom had gained momentum and partially because state power had become strong enough to tax its subjects without the intermediary of lords and vassals.

The phenomenon just discussed throws light on a paradox that has played a role in the recent debate of the transformation of the feudal societies. Hatcher has shown that free peasants were often economically more burdened than the ones with 'unfree' status (1981). The reason is that the latter were not fully exposed to the market forces; they were to some degree protected by the inertia of feudal custom. The French historian R. Boutruche makes a similar observation for late thirteenth-century France: freedom but at a high cost (1970, p. 87).[6]

In the traditional interpretation of the property-relations approach it is the high degree of exploitation allegedly imposed by the feudal bonds that accounts for the technological retardation of mediaeval agriculture. Intuitively, it makes sense that a high degree of exploitation would create problems for technological change at least inasmuch as innovations need resources to get implemented. But if this is the argument, then a loosening of feudal bonds which is accompanied by an increase in exploitation should not stimulate technological progress.

There is, however, a solution to the seemingly contradictory nature of the contending arguments: property relations are decisive but the distribution of income is not. In the model discussed here a fairly unimpressive role is attributed to the degree of exploitation as such. Changes in income distribution will not affect the long-run course of the economy. The decisive factor is the strength of the technological-change parameter and the strength of diminishing returns in agriculture. An increase in the former will foster self-sustained growth and an aggravation of the latter stagnation in per-capita income, but not necessarily at or close to the subsistence level. The importance of property relations thus lies in their impact on the technological-change parameter. Free peasants might pay higher rents and still be technologically more progressive and enjoy higher *net* incomes.

The new institutions that replaced the feudal bonds created incentives for peasants to improve and increase production. Peasants either paid fixed rents (in nominal or real or a combination of nominal and real terms) or entered into some sort of income-sharing system – i.e. crop-sharing. In the latter, both owners of land and producers had an interest in making investment and in improving production.[7] In northern Italy, for example, tenant farmers that made investment in the farm were entitled to legal protection so that they could get compensation if evicted (Miani, 1964; Imberciadori, 1980, p. 442). If the proprietor did not reimburse farmers for their investment then he could not terminate the lease contract. Partly because of this, large landowners, the Church and nobility gradually lost control of the land as they could not reimburse farmers for their investment in improvements (Cipolla, 1947, p. 319).

Improvement in legal status was in itself an incentive – or at least took away a disincentive effect – to invest, and effort and increased freedom of action also made it possible for peasants to take part in the increased specialization, responding more quickly to market opportunities. When peasants gained the freedom to dispose of their own labour it seems as if the hours of work increased as well. This probably contributed to the more advanced crop rotation systems and the attempts to regenerate the fertility of the soil which required a high input of labour. In sum, the changes in the property relations taking place in the twelfth to fourteenth centuries were decisive for the continuation of a dynamic phase in these regions because they gave increased scope for peasant households to increase effort and investments and for the landowners, urban and rural, to invest as well.

At the heart of this argument lies the idea that there are no particular advantages of demesne production from an efficiency point of view. The rationale for the manor was not its efficiency but rather that it was a way for lords to secure a rent from land when it was in abundant supply. This is a controversial argument and it cannot easily be settled because most of our information stems from demesne records and very little from peasant holdings. Theoretical considerations point, however, to the problems of monitoring labour that has low commitment to the goals professed by managers. Work effort on own holdings will, on the other hand, be self-policing. It can also be expected that when farming techniques relied more heavily on labour input as land became scarce, the disadvantages of demesne production would be aggravated. There is also evidence that all of the innovations we have been discussing so far in terms of farming methods were practised by tenants and peasants although available data cannot confirm that these practices originated among them. There are indications that the suppression of the fallow and the introduction of the leguminous crops took place without the consent of landowners. This might be interpreted as evidence of peasants being the innovators (Derville, 1978, 1982, p. 693).

There is – as should be expected from the general line of argument pursued here – evidence of the vigour of small- and middle-sized peasant households. In the Low Countries small holdings are reported to have had higher yields than larger estates. The introduction of horses as a source of traction power in mediaeval England has been the subject of a thorough analysis by J. Langdon, and it is unambiguously clear that the adoption of the horse was more swift among tenants and peasants (1986, e.g. pp. 188–9, 192). One of the advantages of the horse was that it increased the speed of transport and ploughing. The latter released labour for other purposes and the former facilitated market transactions for peasants (p. 271). There is also evidence (on the continent) of more intense use of the labour at small holdings, and where holdings were too small to support a household, nearby cities or larger farms provided opportunities for work in non-peak periods.

3.10 Demographic Regimes and Economic Change

We have argued that the hypothesis of a general crisis in the early fourteenth century caused by overpopulation cannot be upheld. The basic premise of the hypothesis is that technology could not be transformed quickly enough to increase production in line with population. Major economic regions of Europe show evidence to the contrary. Furthermore, if others followed suit a much larger population could obviously be adequately fed. Nor is it correct that intensive land use necessarily led to an exhaustion of the soil. European economic history does not show a uniform pattern. Several independent and largely self-centred urban-agrarian complexes were dynamic in the sense that new plants and methods were introduced and they entertained close relations with the urban economy. Other areas which were socially as well as technologically less modernized seemed to experience a slowdown in population growth and a decline in income. But that does not mean that these parts were slowly moving towards a demographic catastrophe. As has been repeatedly pointed out, Malthusian conditions are consistent with per-capita incomes above subsistence if there is technological progress. Changes in population in the second half of the fourteenth century were dominated by high mortality due to repeated spells of pestilence. Although this demographic sequence had far-reaching economic implications it was largely non-economic in origin.

Recent research in population dynamics indicates that mortality seldom acts as a Malthusian positive check. In pre-industrial economies that adopt the so-called (north) European marriage pattern, economic conditions affect population growth through fertility rather than mortality (see Smith, 1983). Mortality is largely independent of economic conditions; it is primarily exogenously determined, except in periods of severe and prolonged famines. The way that economic conditions determine population growth is through fertility. More specifically, we can find a causal link between real income (which is related to the quality and availability of land) and nuptiality, and between nuptiality and fertility. This evidence then shows that the population is regulated by preventive checks rather than mortality checks to population. A shortage of land increases the age at which couples marry and diminishes the proportion of an age group marrying at all, and both these effects reduce fertility.[8]

The troubled century or so from 1348 was a period when population changes were governed by a cycle of severe epidemics and infectious diseases. Given the fact that this cycle was unrelated to economic conditions we can understand its prolonged nature. If instead it had been a Malthusian positive check, we would have expected an increase in productivity and income and renewed population growth following the first epidemic in 1348. The one-sided emphasis of the demographic approach on the man/land ratio simply cannot explain the prolonged agrarian crisis from the middle of the fourteenth to the end of the fifteenth century.

The present model on the other hand can accommodate the prolonged crisis. A declining or stagnating population will have several contracting effects on the economy. In the first place the criticism spelled out against the simplified ranking of land quality leads us to the pessimistic view about the immediate effects of more and allegedly better land being available. Some marginal land of inferior quality was no doubt abandoned but insofar as abandonment of marginal land was followed by a de-specialization, the compound effect on land productivity might be small or even negative. Yield ratios expressed in terms of seed corn or unit of land, although not easily interpreted, do not exhibit an increasing trend which would be expected from the demographic approach. There is in fact evidence to the contrary showing that yields per acre and seed corn peaked in the second quarter of the fourteenth century (Campbell, 1987). The most obvious interpretation of these findings is that the labour input in production declined in the second half of the fourteenth century.

Even if there was an increase in the (per-capita) supply of land, this increase could not prevent the decline in aggregate income and in aggregate demand for urban products. Urban production having increasing returns would therefore experience a decline in labour productivity. A decline in aggregate demand would increase the relative prices of urban goods which implies that the average peasant household would get a smaller amount of urban goods from a given amount of agrarian goods.[9] Unless the effects of diminishing returns boosted productivity considerably, which we believe is fairly unlikely for reasons spelled out above, the amount of investment goods per peasant household would not improve. Faced with adverse price movements but less expensive land, peasants may also have chosen to diminish their work effort and rely on land-intensive farming. Furthermore, as was stressed in chapters 1 and 2, technological change in pre-industrial economies is based on cultural continuity and inter-generational transmission of a technological heritage. That process was severely disrupted when villages were deserted and families lost those to learn from or those to teach, not only by the repeated epidemics but by wars as well. The sort of self-sustaining growth process typical of the pre-plague era never gets started because it is interrupted by repeated spells of population decline.

3.11 Conclusion

The main arguments in this chapter constitute a critique of a Malthus–Ricardo interpretation of mediaeval European history. That challenge is pursued on both a theoretical and an empirical level. It is argued that population growth endogenously generated factors that countervailed diminishing returns in agriculture. Population pressures tended to loosen the feudal bonds because lords, assisted by market forces that enriched the landowners, found it in their own interest to accept peasants' demands to commute labour services to rents in kind or in money. Freeholders, share-croppers and tenants that gained greater freedom and enjoyed burgeoning demand from a growing urban population created more efficient production units that were responsive to demand conditions and were technologically adaptive. Land scarcity improved the land use not only by an intensification through a suppression of the fallow, but also through new rotation systems and manuring, high labour input in the working of the land, and finally by means of a regional specialization. An increased aggregate demand from a growing population raised the level of productivity in the urban sector. Peasant households could invest in better agricultural implements which raised their productivity as well. Although the whole of Europe did not experience economic progress, those regions that did are important enough to merit attention in a historiography that has been for too long dominated by accounts which have not been able to discern the regional diversity.

NOTES

1 See Desai (1977) for a discussion of the Ricardian argument and a critical assessment of the statistical basis for the belief that corn yields and land productivity were declining in the thirteenth and early fourteenth centuries.
2 This line of argument is obviously based on the implicit assumption that there are no economies of scale in agriculture, which is generally considered a plausible assumption. If there were economies of scale then an agricultural entrepreneur would be able to achieve a higher average productivity by hiring farm labour compared to the productivity of a peasant working his own land. The entrepreneur could consequently earn a rent on land and still pay the farm labourers their opportunity income – i.e. the income earned as free peasants owning their land. If, however, constant returns prevail and peasants are free, then the spontaneously emerging production unit will be the peasant household.
3 Lords may have tried to increase population growth deliberately because of the shortage of labour although direct evidence is limited on this subject.
4 A. Kussmaul's (1981) analysis of servants in early modern England illuminates this aspect. Although her study covers a different period to the one discussed in this chapter there are good reasons to believe that mediaeval servants performed similar roles. See especially chapters 2 and 3–5.
5 G. Bois (1976) is often quoted as having shown that population declined in Normandy in the second quarter of the fourteenth century. Closer scrutiny of his evidence makes such a conclusion premature, however. What Bois does is to infer population changes from two distinct, but neighbouring areas for which there is information about taxable units in the middle of the thirteenth century and in 1314 for one of the areas (let us call it x), and 1347 in the other (to be called z). Assuming that these two areas are representative and identical in all relevant respects and that the proportion of non-taxable units in the population does not change, one can deduce changes in population by comparing the population indices for the two areas. An index is the ratio (taxable units in area x in 1314)/(taxable units in area x c. 1250) and is reported as being slightly larger than the ratio (taxable units in area z in 1346)/(taxable units in area z c. 1250). The two areas are not identical in all relevant respects, however. Area x has a much larger population, a larger sample of villages and a larger average village size, and furthermore an untypically large village (Auffay) accounting for 15 per cent of the total and almost ten times the average size of the villages in sample z. Finally, that particular village has above average population growth up to 1314. If we omit that particular village results are reversed. Instead of a slight decline in population between 1314 and 1346 there is a slight increase!
6 It is not clear that the Hatcher paradox is generally valid, and no such claim of generality has been advanced. In the Low Countries and in some parts of Italy it seems as if owners of land showed a certain restraint in their exaction of rents possibly because they know that there was a long-term interest in having a stable relationship with tenants.
7 As demonstrated by Braverman and Stiglitz (1986) landowners may find it in their

own interest to resist innovations preferred by tenants. The specific characteristic of such a technique is that, although superior, it has diminishing returns at lower levels of effort than the inferior technique. I find that characteristic fairly untypical for the type of technological changes that we are considering in this period, however.
8 There is some controversy as to the way in which, and with what lag, economic conditions affect population growth. It is possible that population growth may be related to the *change* in per-capita income rather than its level. That would make it possible to reconcile findings of decline in population growth with stagnating but comparatively high income.
9 Furthermore, the higher mortality in cities could lead to an excess demand for urban goods explaining, in conjunction with the declining urban productivity, the relative decline of agrarian prices that can be observed after some time. Urban prices and incomes thus go up until migration has again levelled relative incomes. But since there is a prolonged cycle of recurrent epidemics that levelling process takes time. The decline of agrarian prices in terms of urban products seems to be recorded in most parts of Europe and terms of trade remain in that state during most of the fifteenth century (Abel, 1980a, pp. 49–53).

APPENDIX   by Peter Skott

Production

Apart from land, which is in fixed supply, there are two inputs, labour (L) and capital (K). The capital stock changes over time as a result of investment and depreciation. If I denotes gross investment and δ is the rate of depreciation then the rate of change of the capital stock is given by

$$\dot{K} = I - \delta K, \tag{A3.1}$$

where a dot over a variable is used to denote the rate of change.

Labour input is measured in 'efficiency units': the higher are the real incomes of peasants, the greater will be the efficiency input, E_a,

$$E_a = L_a f(y_a), \tag{A3.2}$$

where y_a is an index of per-capita real income of peasants (to be defined below). For simplicity, we let the function $f(\cdot)$ take the log-linear form

$$f(y_a) = y_a^\varepsilon; \quad \varepsilon \geq 0. \tag{A3.3}$$

Notice that this specification of efficiency labour may capture two distinct effects. Higher incomes and a more varied consumption pattern lead to increased specialization and a greater division of labour which in turn cause productivity gains. In addition to the specialization effect, there is the normal efficiency-wage effect as better nutritional standards enhance work capacity.

Assuming that the production function is a monotonic transformation of a simple Cobb–Douglas function, and suppressing the fixed input of land, total agricultural output (A) is given by

$$A = F(K^\alpha E_a^{1-\alpha}) e^{rt}, \qquad (A3.4)$$

where r is the rate of disembodied technical progress and where F is monotonically increasing and has elasticity less than or equal to one (i.e. there are diminishing or constant returns to scale).

Urban production is characterized by increasing returns to scale and in addition we allow for exogenous technological progress. For simplicity it is assumed that there is a single input, labour, but that efficiency depends on real incomes, y_u, as in agriculture. Algebraically, we write the production function

$$U = E_u^{1+\beta} e^{\rho t}; \quad \beta \geqslant 0, \qquad (A3.5)$$

where ρ is the rate of disembodied technical progress and

$$E_u = L_u f(y_u) = L_u y_u^\varepsilon, \qquad (A3.6)$$

and y_u is an index of urban per capita income.

Demand

At low income levels almost all income is spent on agricultural produce but as incomes rise, an increasing proportion will be allocated to urban goods. The simplest way to capture this is through a linear expenditure system (corresponding to a Stone–Geary utility function). If $(1-\theta)$ is the share of rent in agriculture then nominal peasant income is $\theta p_a A$. We assume that all income is spent and the demand of peasants for agricultural and urban goods can therefore be written

$$A_p^d = L_a(c_0 + c\theta A/L_a), \qquad (A3.7)$$

$$U_p^d = L_a(-c_0 + (1-c)\theta A/L_a) p_a/p_u, \qquad (A3.8)$$

where the superscript d denotes demand and the subscript p indicates the origin of demand, peasants. In order for U_p^d to be positive (and for peasants' demand for agricultural products to fall below their share of agricultural output) we must have $\theta A > L_a c_0/(1-c)$; we assume that this inequality is satisfied.

Analogously, we have the following urban demand for agricultural and urban goods:

$$A_u^d = L_u(c_0 + cp_u U/(p_a L_u)), \qquad (A3.9)$$

$$U_u^d = L_u(-c_0 p_a/p_u + (1-c)U/L_u), \qquad (A3.10)$$

where, for simplicity, it has been assumed that the demand parameters c_0 and c are the same in the two sectors. The condition for positive urban demand for urban goods is $U > L_u c_0 p_a/((1-c)p_u)$, and we assume that this is satisfied.

In the case of landlords' rent income, the asymmetric demand specification loses its rationale, and instead we assume that a constant proportion of rent is spent on agricultural products and the remainder on urban goods, i.e.

$$A_r^d = \sigma(1-\theta)A, \qquad (A3.11)$$

$$U_r^d = (1-\sigma)(1-\theta)p_a A/p_u. \qquad (A3.12)$$

For expository simplicity we let $\sigma = c$ which does not in any way affect the qualitative results. Combining equations A3.7–A3.12 we find expressions for the total demand for agricultural and urban goods,

$$A^d = c_0 L + A(c\theta + \sigma(1-\theta)) + U c p_u/p_a = c_0 L + cA + cU p_u/p_a \qquad (A3.13)$$

$$U^d = -c_0 L p_a/p_u + A[\theta(1-c) + (1-\sigma)(1-\theta)]p_a/p_u + U(1-c)$$
$$= -c_0 L p_a/p_u + (1-c)A p_a/p_u + U(1-c), \qquad (A3.14)$$

where L is total urban and agricultural employment,

$$L = L_a + L_u. \qquad (A3.15)$$

Prices and Sectoral Output Levels

We assume that prices adjust so as to clear the two product markets, and since Say's law has been imposed, the simultaneous clearing of both markets is ensured if agricultural demand equals agricultural production; i.e. if

$$A = (c_0 L + Ac) + U c p_u/p_a, \qquad (A3.16)$$

or

$$p_u/p_a = [A(1-c)-c_0L]/(cU). \quad (A3.17)$$

We assume that per-capita incomes of workers are equal in the two sectors. Undoubtedly, there were fluctuations in relative incomes during the mediaeval period. However, information on this issue is very limited and there is no evidence of a secular trend in relative incomes. We therefore assume constant relative incomes and for expository simplicity we assume a uniform per-capita income. Urban taxation is ignored and the net income of peasants is described by the distributional parameter θ. Algebraically, the equality between net rural and urban incomes can therefore be written

$$\theta p_a A/L_a = p_u U/L_u, \quad (A3.18)$$

and using equations A3.17 and A3.18 we get

$$\theta(A/L_a)(L_u/U) = [A(1-c)-c_0L]/(cU). \quad (A3.19)$$

If we denote the productivity of labour in agriculture by q_a, $q_a = A/L_a$, this can be rewritten

$$L_a q_a(1-c) - c_0 L = \theta q_a L_u c, \quad (A3.20)$$

or

$$l = L_a/L_u = [c\theta q_a + c_0]/[q_a(1-c)-c_0], \quad (A3.21)$$

where l is the ratio of agricultural to urban employment.

From equation A3.21 we can derive an expression for the change in the sectoral composition of the labour force. Differentiating logarithmically, we get

$$\begin{aligned}\hat{l} &= \hat{L}_a - \hat{L}_u \\ &= -c_0(1-c(1-\theta))/\{[q_a(1-c)-c_0][c\theta q_a+c_0]\}\hat{q}_a \\ &= -\phi(q_a)\hat{q}_a; \quad \phi > 0, \quad \phi' < 0,\end{aligned} \quad (A3.22)$$

where $\hat{\ }$ denotes a proportional growth rate, $\hat{x} = 1/x \, dx/dt$.

Population Growth

Equations A3.21 and A3.22 describe the sectoral composition of workers in agriculture and urban employment. For simplicity it is assumed that the total work force in these two sectors is proportional to the total population. We thus ignore variations in the proportion of the labour force which is employed directly by landlords (soldiers, servants, etc.). Demographic shifts (e.g. changes in the age composition) which may cause changes in the relative size of the working population are also disregarded.

In accordance with the standard Malthusian view, finally, we posit a direct functional relation between real per-capita income and population growth,

$$\hat{L} = (L_a \hat{+} L_u) = n(y); \quad n' > 0, \tag{A3.23}$$

where $y = y_a = y_u$ is the index of real per-capita income. The question now is how to define y.

The Stone–Geary utility function does have a corresponding real-income index, ω.[1] It is not, however, obvious that this 'correct' index is the appropriate one for the purpose at hand. The possession of silver jewellery may yield positive utility without necessarily stimulating agricultural productivity, and the Malthus effect on mortality and fertility rates is not clear either. Instead of the Stone–Geary index we shall therefore use a different and simpler real-income index: we assume that both productivity levels and population growth will be determined by per-capita consumption of agricultural products, and the relevant index therefore becomes

$$y = y_a = y_u = c\theta q_a + c_0. \tag{A3.24}$$

Combining A3.23 and A3.24 we get

$$\hat{L} = \psi(\theta q_a); \quad \psi' > 0. \tag{A3.25}$$

It is reasonable to suppose that $\hat{L}$ is bounded above but not below: although the growth of population can never exceed some given positive number, there is a minimum level of food intake below which survival becomes impossible. We therefore assume that

$$\psi \leqslant n^{\max} \quad \text{for all } \theta \text{ and } q_a,$$
$$\psi \to -\infty \quad \text{for } \theta q_a \to c_a^{\min}, \tag{A3.26}$$

where $c_a^{\min}$ is the minimum survival level. It is convenient (but not necessary) to assume that this level is above the 'critical' value of real income in the consumption function, i.e. that $c_a^{\min} \geq c_0/(l-c)$; this implies that the demand condition $\theta q_a > c_0/(l-c)$ will be fulfilled automatically.

By definition,

$$\hat{L} = L_a/L\,\hat{L}_a + L_u/L\,\hat{L}_u = 1/(1+l)\hat{L}_a + 1/(1+\hat{l})\hat{L}_u, \qquad (A3.27)$$

and using equations A3.23, A3.25 and A3.27, we get expressions for $\hat{L}_a$ and $\hat{L}_u$,

$$\hat{L}_a = \psi(\theta q_a) + 1/(1+l)\hat{l}, \qquad (A3.28)$$
$$\hat{L}_u = \psi(\theta q_a) - 1/(1+\hat{l})\hat{l}, \qquad (A3.29)$$

where l and $\hat{l}$ are given by equations A3.21 and A3.22, respectively.

Investment

In order to complete the model we still need to specify investment levels in agriculture. We assume, somewhat unrealistically, that agricultural capital consists entirely of urban goods and that a constant proportion, τ, of all urban goods acquired by peasants is invested. Notice that investment is undertaken exclusively by peasants and not by landlords. With these assumptions we have (using A3.1 and A3.7),

$$\dot{K} = L_a \tau\, p_a/p_u\,[-c_0 + (l-c)\theta q_a] - \delta K, \qquad (A3.30)$$

or, using equation A3.18,

$$\hat{K} = q_u\,L_a/K\,\tau\,1/(\theta q_a)\,[-c_0 + (l-c)\theta q_a] - \delta. \qquad (A3.31)$$

Analysis

The model can be reduced to a two-dimensional system of differential equations. Define a new variable, x, by

$$x = q_u L_a/K. \qquad (A3.32)$$

The variable x is introduced exclusively for analytical convenience and we have found no clear intuitive interpretation of it. Equations A3.1–31 imply that

$$\hat{q}_a = 1/f(\theta q_a)[-A\psi(\theta q_a) + B(\theta q_a)x - C], \tag{A3.33}$$

$$\hat{x} = [D - E(q_a, \theta q_a) - Ff(\theta q_a)]\hat{q}_a - G\psi(\theta q_a) + H, \tag{A3.34}$$

where

$$\begin{aligned}
f(\theta q_a) &= 1 - \gamma(1-\alpha)\varepsilon - (1-\gamma(1-\alpha)(1+\varepsilon))c_0/(c_0 + c\theta q_a) \\
&= (1-\gamma(1-\alpha)\varepsilon)(1 - c_0/(c_0 + c\theta q_a)) \\
&\quad + \gamma(1-\alpha)c_0/(c_0 + c\theta q_a) > 0, \\
A &= 1 - \gamma(1-\alpha) > 0, \\
B(\theta_a) &= \gamma\alpha\tau(-c_0 + (1-c)\theta q_a)/(\theta q_a) > 0, \\
C &= \gamma\alpha\delta - r, \\
D &= (1+\beta)\varepsilon \geq 0, \\
E(\theta q_a) &= (1+\varepsilon(1+\beta))c_0/(c_0 + c\theta q_a) - \beta c_0/[q_a(1-c) - c_0], \\
F &= 1/(\gamma\alpha) > 0, \\
G &= (1-\gamma(1-\alpha))/(\gamma\alpha) - (1+\beta), \\
H &= \rho + r/(\gamma\alpha) > 0,
\end{aligned}$$

and where γ is the elasticity of F; γ is thus the returns-to-scale parameter and may not remain constant over time.

Consider the existence of equilibrium. If $\hat{q}_a = \hat{x} = 0$, then

$$\psi(\theta q_a^*) = (r + \gamma\alpha\rho)/[1 - \gamma(1+\alpha\beta)], \tag{A3.35}$$

$$\begin{aligned}
x^* &= [\rho + \delta + (1+\beta)(r + \gamma\alpha\rho)/(1-\gamma(1+\alpha\beta))] \\
&\quad \div [\tau(-c_0 + (1-c)\theta q_a^*)] \theta q_a^*,
\end{aligned} \tag{A3.36}$$

and an equilibrium thus exists if

$$\lim_{\theta q_a \to \infty} \psi(\theta q_a) > (r + \gamma\alpha\rho)/[1 - \gamma(1+\alpha\beta)], \tag{A3.37}$$

$$x^* > 0. \tag{A3.38}$$

Assuming that an equilibrium exists, the dynamics of the system in the neighbourhood of equilibrium is determined by the Jacobian of the system

$$J(q_a^*, x^*) = \begin{Bmatrix} q_a^* a & q_a^* b \\ -x^*(ka+d) & -x^* kb \end{Bmatrix}, \tag{A3.39}$$

where

$$a = \partial \hat{q}_a/\partial q_a = 1/\mathrm{f}(\theta q_a)\,[\gamma\alpha\tau c_0/(\theta q_a^2)x - (1-\gamma(1-\alpha))\theta\psi'(\theta q_a)],$$

$$b = \partial \hat{q}_a/\partial x = 1/\mathrm{f}(\theta q_a)\,\gamma\alpha\tau(-c_0 + (1-c)\theta q_a)/(\theta q_a)$$
$$= B(\theta q_a)/\mathrm{f}(\theta q_a) > 0,$$

$$k = -(\partial \hat{x}/\partial x)/(\partial \hat{q}_a/\partial q_a) = 1/(\gamma\alpha)\,\mathrm{f}(\theta q_a) - (1+\beta)\varepsilon + (1+\varepsilon(1+\beta))c_0$$
$$\div (c_0 + c\theta q_a) - \beta c_0/[q_a(1-c) - c_0] = F\mathrm{f}(\theta q_a) - D + E(\theta q_a),$$

$$d = -(\partial \hat{x}/\partial q_a + ca) = [(1-\gamma(1-\alpha))/(\gamma\alpha) - (1+\beta)]\theta\psi'(\theta q_a)$$
$$= G\theta\psi'(\theta q_a).$$

The determinant and trace of J are given by

$$\mathrm{Det}(J) = x^* d q_a^* b, \tag{A3.40}$$

and

$$\begin{aligned}\mathrm{TR}(J) &= q_a^* a - x^* k b \\ &= 1/\mathrm{f}(\theta q_a)\,[\gamma\alpha\tau c_0/(\theta q_a)\,x^* - (1-\gamma(1-\alpha))q_a\theta\psi' \\ &\quad - k\gamma\alpha(\rho+\delta+(1+\beta))\psi^*] \\ &< \gamma\alpha(\rho+\delta+(1+\beta))\psi^*/\mathrm{f}(\theta q_a) \\ &\quad \times [c_0/(-c_0 + (1-c)\theta q_a) - k].\end{aligned} \tag{A3.41}$$

The equilibrium is a saddlepoint if the determinant of J is negative and an unstable node/focus if Det > 0 and TR > 0. Stability ensues if Det > 0 and TR < 0.

When the equilibrium is stable it may be of interest to examine the comparative static effects of changes in the parameters. Table A3.1 sets out the results for the variables and parameters of primary interest. Assuming stability, we have $1 - \gamma(1+\alpha\beta) > 0$ and the equilibrium rate of population growth is, as indicated in column 1, positively related to the technological progress parameters (r and ρ), to the scale parameters (γ and β) as well as to the capital coefficient (α). The reasons for the direct relationship with r, ρ, γ, β are straightforward and need no comment. The positive effect on ψ^* of a rise in α is due to the fact that a greater weight is thereby attached to capital in the production of agricultural output, and the capital good itself is produced under increasing returns.

Population growth is an increasing function of per-capita con-

TABLE A3.1 Effects of changes in parameters

	ψ^*	y^*	x^*	l^*
r	$1/(1-\gamma(1+\alpha\beta))$	$1/\psi'\,\partial\psi^*/\partial r$	$A\,\partial\psi^*/\partial r$	$B\,\partial y^*/\partial r$
ρ	$\gamma\alpha/(1-\gamma(1+\alpha\beta))$	$1/\psi'\,\partial\psi^*/\partial\rho$	$A\,\partial\psi^*/\partial\rho + 1/[\tau(-c_0/(\theta q_a^*)+1-c)]$	$B\,\partial y^*/\partial\rho$
γ	$(r+\alpha\gamma+\alpha\beta r)/[1-\gamma(1+\alpha\beta)]^2$	$1/\psi'\,\partial\psi^*/\partial\gamma$	$A\,\partial\psi^*/\partial\gamma$	$B\,\partial y^*/\partial\gamma$
α	$\gamma(\rho(1-\gamma)+\beta r)/[1-\gamma(1+\alpha\beta)]^2$	$1/\psi'\,\partial\psi^*/\partial\alpha$	$A\,\partial\psi^*/\partial\alpha$	$B\,\partial y^*/\partial\alpha$
β	$\gamma\alpha(r+\gamma\alpha\rho)/[1-\gamma(1+\alpha\beta)]^2$	$1/\psi'\,\partial\psi^*/\partial\beta$	$\psi^*[\tau(-c_0/(\theta q_a^*)+1-c)] + A\,\partial\psi^*/\partial\beta$	$B\,\partial y^*/\partial\beta$
θ	0	0	0	$l^* q_a^*[(1-c)/\theta]$
ε	0	0	0	0
τ	0	0	$-x^*/\tau$	0
δ	0	0	$1/[-(c_0/(\theta q_a^*)+1-c)]$	0
c	0	0	$x^*[c_0/(c\theta q_a^*)+1][1-c-c_0/(\theta q_a^*)]$	$l^* q_a^*[(1/c)]C$
c_0	0	0	$x^*[1+c_0/(c\theta q_a^*)]/[(1-c)\theta q_a^* - c_0]$	$l^*[1+(1-c)/c\theta]C$

where
$A = [1+\beta-x^*\tau c_0/(c(\theta q_a^*)^2\psi')]/[1-c-\tau c_0/(\theta q_a^*)]$,
$B = [1-l^*(1-c\theta-\sigma(1-\theta))/(c\theta)]/[q_a^*(1-c)-c_0]$,
$C = 1/[q_a^*(1-c)-c_0]$.

sumption of agricultural products, $\psi = \psi(y)$ and hence $dy = (1/\psi')d\psi$. It therefore follows (column 2) that this income/consumption measure (y) is also positively related to $r, \rho, \gamma, \beta, \alpha$.

But perhaps somewhat surprisingly, neither ψ^* nor y^* are affected by changes in other parameters. In fact, the parameters θ and ε, so crucial to the property-relations approach and the commercialization approach, appear to have no impact on x^* either. Why does a change in θ have no effect on the long-run equilibrium outcome? If θ is raised then the initial impact is to improve the standard of living of peasants. This improvement causes the growth rates of both the population and the capital stock to go up, but because of diminishing returns this gradually forces down real incomes until a new long-run equilibrium is reached; the new equilibrium values of ψ and x being identical to the old ones. The effect of a change in θ is a one-off change in the size of the population, the long-term growth rate being unaffected.[2]

The commercialization parameter, ε, is somewhat different. Changes in ε have two distinct effects. From equations A3.2, A3.3, A3.5 and A3.6 it follows that an increase in ε will lead to a rise (fall) in work intensity if $y > 1$ ($y < 1$). This change in intensity causes real per-capita incomes to change in exactly the same way as following a change in θ, and it gives rise to similar (positive or negative) level effects on population and the capital stock. But these level effects are not really related to the *raison d'être* of including ε as a parameter. We included ε because it seemed reasonable to posit a non-zero elasticity of work intensity with respect to changes in income. But since real per-capita incomes are constant in long-run equilibrium, a change in the elasticity has no effects at all on the equilibrium configuration.

Changes in the remaining parameters, δ, τ, c, c_0, also fail to have permanent growth effects on ψ^*. Like θ and ε they have level effects but, unlike in the case of θ and ε, the level effects are 'unbalanced': a change in any of these parameters will affect $x^* = L_a q_u/K$ (see column 3) and thus, given productivity in urban production, the labour/capital ratio in agriculture.

The sectoral composition of labour is also constant in long-run equilibrium, and whereas changes in the demand parameters, c, c_0, had no effect on the growth rate, they exert a powerful influence on l^*. This is not surprising. The two parameters determine how a given income is allocated between the consumption of agricultural and urban goods. An increase in either of the parameters will raise the share of expenditure on agricultural produce and thus the equilibrium ratio of peasants to urban workers. It should be noted, however, that given the specification of the 'Malthus function', the per-capita consumption of agricultural produce is invariant with respect to changes in c and $c_0 : \psi^*$ is constant and $\psi^* = \psi^*(c_0 + c\theta q_a^*)$. Changes in c and c_0 therefore influence relative sectoral demands entirely through their effects on the demand for urban goods.

The other parameters have no direct influence on l^*, but since the real income level, y, affects the share of agricultural produce in peasant and worker consumption, all the variables affecting y^* will have an indirect effect on l^*.

Other variables, apart from ψ^*, y^*, x^* and l^*, may also be of interest. What happens, for instance, to relative prices, urban productivity and the capital–output ratio in agriculture? None of these variables will be constant over time. From equations A3.5 and A3.6 we know that

$$q_u = U/L_u = L_u^\beta y_u^{\varepsilon(1+\beta)} e^{\rho t}, \qquad (A3.42)$$

and hence

$$\hat{q}_u = \beta \hat{L}_u + \varepsilon(1+\beta)\hat{y}_u + \rho, \qquad (A3.43)$$

and if $\hat{y}_u = 0$,

$$\hat{q}_u = \beta \psi^* + \rho. \qquad (A3.44)$$

In equilibrium, urban productivity will thus be growing exponentially at the rate $\beta \psi^* + \rho$. Since $p_a \theta q_a = p_u q_u$ and θq_a is constant in equilibrium, this implies that

$$\hat{p}_a - \hat{p}_u = \hat{q}_u = \beta \psi^* + \rho. \qquad (A3.45)$$

The ratio of agricultural to urban prices will thus be growing at the same rate as urban productivity, and indeed so will the capital–labour ratio in agriculture: since $x = L_a q_u / K$ we have

$$\hat{K} - \hat{L}_a = \hat{q}_u - \hat{x}, \qquad (A3.46)$$

and in equilibrium $\hat{x} = 0$.

Implications

In this section we examine the implications of the model for the debate on mediaeval growth. Initially, during the early part of the period, land was (relatively) abundant and there may have been near-constant returns to scale in agriculture, i.e. $\gamma \simeq 1$. Equations A3.35–38 show that under these conditions an equilibrium may not exist or, if it does, it may be a saddlepoint. If $\psi(z) \to n^{\max}$ for $z \to \infty$ then it follows from A3.37 that there is no equilibrium for

$$M \leqslant \gamma \leqslant N, \tag{A3.47}$$

and that the equilibrium is a saddlepoint for

$$\gamma > N, \tag{A3.48}$$

where

$$M = [n^{\max} - r]/[\alpha\rho + n^{\max}(1+\alpha\beta)], \text{ and}$$
$$N = [1 + (r(1+\beta) + \rho)/\delta]/[1 + \alpha\beta + \rho(1-\alpha)/\delta].$$

In the case where $M \leqslant \gamma \leqslant N$, the economy will exhibit sustained growth in real per-capita income. Intuitively, this is not surprising. If the degree of diminishing returns in agriculture is small relative to the rates of technical progress and the degree of increasing returns in urban production, then it is impossible to increase the population fast enough to keep real incomes down.

If γ is increased further and $\gamma > N$ then an equilibrium exists but it will be a saddlepoint and the economy will almost certainly diverge from the equilibrium. The direction of divergence depends on the initial configuration; high initial values of θq_a and x will lead to perpetual expansion whereas low starting values will cause contraction. The intuition behind unlimited expansion should be clear but how can we get contraction? If, loosely speaking, the combined effect of increasing returns in urban production and weakly diminishing or constant (or increasing) returns in agriculture is to make the overall system subject to increasing returns, then low real incomes and the associated decline in population will cause productivity to fall and thus induce a further decline in real incomes.

How high will γ need to be before an equilibrium ceases to exist? The answer obviously depends on the value of other parameters but for all reasonable assumptions the critical value, M, will be somewhere between 0.9 and 1.[3] Whether or not the returns to scale in agriculture were so close to unity in the early part of the period is an open question. There is little doubt, however, that the condition was close to being satisfied in many regions in the period after the Black Death; it is hard to see how returns to scale could be anything but close to one following the decimation of the population by 25–50 per cent.

Matters are different for the period immediately preceding the Black Death. According to the demographic approach the doubling of population over the eleventh and twelfth centuries meant that by the thirteenth century there were strongly diminishing returns[4] and this, it is argued, led to stagnation and crisis. Does the model support this argument? If γ is below the critical value (M) the economy settles in an equilibrium that is either a node or a focus and it is readily seen (using equation A3.41) that if γ is low enough, then the equilibrium must be stable; for reasonable parameter values the equilibrium will in fact be stable for all $\gamma < M$. Since equilibrium real income and population growth rates are both positively related to γ it follows that a decline in γ will lead to stagnation and falling real income; the demographic scenario is apparently born out by the model.

The argument against the demographic explanation focuses on four different aspects of the model:

efficiency effects of increasing specialization (the parameter $\varepsilon \geqslant 0$);
the distribution of income (θ describing the share of peasants);
increasing returns in urban production ($\beta \geqslant 0$); and
technical progress (r and $\rho \geqslant 0$).

With respect to the first two factors, the striking implication of the model is that in equilibrium neither real income nor population growth depend on ε and θ (cf. equations A3.35–6). This does not mean that changes in ε and θ have no effects at all on the outcome. Changes in the two parameters have temporary effects on net income and population growth. An increase in the share of peasants, for instance, will delay the depressing effects of a fall in γ. In this sense the property approach is right in arguing that shifts in the distribution of income may offset the effects of diminishing returns. Proponents of the approach should, however, add that the respite is temporary and they may also find it difficult to explain why there should be a significant distributional shift *in favour of peasants* at a time of increasing pressure on land. Insofar as criticisms of the demographic approach focus on distributional aspects alone, they thus do not seem compelling.

The critics are on more solid ground when they point to endogenous effects of population growth on technical progress and increasing returns. If the increasing scarcity of land leads to a loosening of feudal bonds and thereby raises the rate of technological progress (r and ρ) and increases the scope for specialization (i.e. raises β) then the effect on income and growth becomes indeterminate (equations A3.35–6). As the density of population increased in the eleventh to thirteenth centuries, the elasticity of agricultural output with respect to reproducible inputs may have declined, but if that decline was associated with faster technical progress then the outcome may have been a steady rise in equilibrium net income and in the rate of population growth. The outcome clearly depends on the relative strength of the different effects, a question which *a priori* reasoning cannot answer.

NOTES

1 The real per-capita income measure associated with the Stone–Geary utility function is

$$\omega = \begin{cases} \theta q_a & \text{for } p_a \theta q \leq c_0/(1-c), \\ c_0/(1-c) + p_a(\theta q_a - c_0/(1-c))/(p_a^c p_u^{1-c}) & \text{otherwise.} \end{cases}$$

Notice that within a Malthusian framework it would not be in the long-run interest of landlords to squeeze the share parameter, θ, as much as possible even if they were able to do so. If they wish to maximize the absolute amount of rent then the negative effect of a decline in the share of peasants on the size of peasant population should be taken into account. In a long-run Malthusian equilibrium the real per-capita income of peasants is independent of the share parameter.

2 The critical value is given by $M = [n^{max} - r]/[\alpha\rho + n^{max}(1+\alpha\beta)]$ and if

$2\% \geqslant n^{max} \geqslant 1\%$
$0.1\% \geqslant r \simeq \rho \geqslant 0.02\%$,
$0.1 \geqslant \alpha$,
$0.1 \geqslant \beta \geqslant 0$,
then it follows that
$99/100 \geqslant M \geqslant 90/102$.

3 This hypothesis is also consistent with the Domar view of the loosening of feudal bonds.

4 The available evidence clearly suggests that θ was less than 3/4 and that peasants' demand for urban goods was at least as large as landlords' demand for urban goods. If we combine these two 'stylized facts' with the reasonable assumption that $\sigma \leqslant c$ then $c_0/(-c_0 + (1-c)\theta q_a)$ is less than 2. If $c \simeq 9/10$ then the value of k on the other hand cannot fall below 7/2 as long as $\gamma \leqslant 9/10$ and $\alpha \leqslant 1/4$. The trace is thus negative and the equilibrium is (locally asymptotically) stable.

4 Measuring the Immeasurable: Labour Productivity in the European Mediaeval Economy

4.1 Introduction

The case argued in the preceding chapters for the existence of periods of substantial growth of labour productivity in pre-industrial times has been considered unconvincing by many historians because of the lack of clear evidence. There are, of course, no time-series data over sectoral inputs and outputs or aggregate production figures. From the thirteenth century onwards there are, however, an increasing number of manorial and monasterial records that give indications, albeit partial, of trends in production, population, rent and crops grown. The painstaking efforts of mediaeval historians to analyse these records have greatly increased our knowledge of European history. But this material has severe deficiencies when it comes to using it for the important task of measuring productivity let alone the estimation of welfare.

The production units covered by these records are the manors and their peasants. But, as argued in chapter 3, the innovative centre shifted from manors to independent peasants and tenants as the de-feudalization process gained momentum. If that is true we are left with the suspicion that these records tell us more about manorial (mis)management than about mediaeval agriculture in general.[1]

One important source of production trends is the tithe statistics. Having first attracted systematic interest in France among historians such as E. Labrousse, J. Goy and E. Le Roy Ladurie they later generated a multinational research project. Some national reports have already been published.[2] Notwithstanding the term, the tithe was seldom exactly ten per cent of the produce. In fact it varied between five and 15 per cent of the physical products from agriculture. Not only the changing incidence of fraud but also the difficulties in making a correct estimation of the percentage officially taken by the tithe, both of which reflected the social climate and fluctuated, will affect the production figure derived from the observed physical size of the tithe. The volume of, for example, wheat produced is calculated on the basis of a supposed tithe, say ten per cent of total wheat production, making total production ten times the tithe. A one percentage point misspecification of the proportion taken by the tithe will consequently result in a ten per cent error in the estimation of the total output. This calls for some caution and skill in the interpretation of the production figures derived from the tithe statistics. A margin of error of up to ± 25 per cent has been suggested.

Once series of production of agricultural products have been established, additional problems arise if they are used as a basis for estimations of productivity. In that case, production must be calculated in terms of a known input of land or labour. Tithe statistics have been used for calculations of land productivity. Calculation of land productivity nonetheless creates serious problems, especially in the analysis of long-term trends, because it is very difficult to control for changes in the amount of land used for agricultural purposes. The difficulties do not end here, unfortunately, since the material is far from complete in the recording of agricultural products actually produced. Information is fairly good as far as corn and to some extent vegetables are concerned but less reliable, or lacking altogether, when it comes to wine and livestock. Tithe statistics will not, in other words, provide an unproblematic basis for estimation of aggregate agrarian output. It is thus 'hypothetical to extrapolate output from tithe statistics' as one scholar reminds us (cf. Daelemans, 1982, p. 30), let alone land productivity.

Manorial records also provide a basis for measurement of productivity of land and seed corn, so-called yield ratios. There is a longstanding tradition among mediaeval historians of discussing the evolution of harvested corn per seed corn or per unit of land.[3] This type of data is not available for all types of arable products; we lack information about the industrial plants which became increasingly important in the last centuries of the mediaeval era. Changes in quality are also ignored as well as improvements in storage techniques. Apart from this, the yield ratios are not easily interpreted unless the density of sowing is also taken into consideration. An increase in the density of sowing will, other things being equal, decrease the yield ratio per seed corn but presumably increase the yield per unit of land. Other things are not always equal, however. Sometimes high seeding rates are but an aspect of labour-intensive or efficient farming methods so that high density and high yields per seed corn are correlated on some holdings. In periods of land scarcity, peasants may choose to increase the density of sowing as an appropriate method to increase production. And, conversely, when prices are declining the chosen strategy may be a deliberate contraction of production by sparse sowing, decreasing the yield per unit of land especially if input of labour and manure decreases as well.[4]

Another problem in the interpretation of yield data is the fact that the frequency of cropping may differ and change over time. B. Campbell's analysis of Norfolk agriculture takes different fallowing periods into consideration with the result that the overall productivity of land over an extended agricultural cycle improved as a consequence of the reduced fallow. This result is not surprising but has not been discovered until recently (1983a, b). More unexpectedly, yield ratios calculated per seed corn did not seem to be negatively affected by frequent cropping. But since most studies do not control for the intensity of land use there is an obvious difficulty involved in interpreting the yield data.

The difficulties encountered in establishing and in interpreting the traditional productivity measures can be summarized as follows:

1 The material is usually partial, based on manorial or monasterial units. The representativity of such units may diminish over time.

2 There is an aggregation problem in two respects. Firstly, because important agrarian goods are included neither in yield nor tithe statistics. Secondly, because there are difficulties in aggregating a number of products expressed in physical units to a compound agrarian product or income and consequently to measure productivity, i.e. changes in that entity with respect to an input, say, land or labour.

3 Changes in composition of the aggregate product, including the introduction of new goods, cannot be reliably estimated, although such changes have important repercussions for the welfare of the population.

4 Changes in the quality of products cannot be taken into consideration when physical measures such as weight or volume are used.

5 Changes in technology and productivity related to the processing of goods, rather than the actual cultivation, such as milling, transport and storing are outside the realm of productivity measurement based on tithe or yield statistics which are based on (part of) the cultivated output. Such improvements have repercussions on welfare, however.

6 The measures used, being related to the productivity of land or seed corn, do not give appropriate information about the welfare aspects of changes in productivity. A change in land productivity cannot unambiguously be interpreted as a corresponding change of income because, for example, a decline in the yield per unit of land can be part of an income-maximizing strategy.

These points suggest that although mediaevalists have made extremely important contributions in their analysis of yield and tithe data it would be highly desirable if an alternative and complementary approach could be found that is immune to the above-mentioned criticisms. It is the purpose of this chapter to provide such an approach. The plan of the remaining part of the chapter is as follows. In section 4.2, the basic ideas as well as the type of empirical information needed for this alternative approach to productivity measurement are presented. In section 4.3, some tentative results are presented with regard to European mediaeval history. The formal structure of the method used is described more rigorously in the appendix.

4.2 Towards a New Approach

The great advantage of the present method for measuring labour productivity in the agrarian sector is that it needs comparatively limited amounts of empirical information. That is not to say that the required information is easily available or of a quality that makes the calculations immune to criticism. But the critique that may be raised against this method will be different from the critical remarks made against the traditional procedures based on yield or tithe statistics.

The purpose of this chapter is to a large extent methodological. The intention is to show the possibility of an alternative solution to the puzzling problem of productivity measurement when only indirect evidence is available, a predicament for the analysis of all pre-industrial economies. The neatness of the solution suggested here will hopefully stimulate other historians to improve the empirical basis so that eventually more robust estimations can be produced. Needless to say, the particular results that are presented below, plausible as they may appear, are not the unique outcome of this method. The method is general enough to accommodate any reasonable amendment of the empirical information. The generality of the method also makes it possible to use it on other historical epochs and areas, for example the European antiquity and for those periods of Chinese economic history for which the limited amount of information needed is available.

The structure of the method employed in this chapter is related to the model discussed in the preceding chapter and was briefly discussed there. A short recapitulation may be necessary, however. In a closed economy in which there are two types of goods, agrarian and urban goods, the agrarian population produces all the food for the entire population. Part of that production is exchanged for urban goods manufactured by urban producers, part of it is expropriated by the land-holding classes using it for their own consumption and to buy urban goods. Assume now that agrarian labour productivity is increasing and that the product expropriated by the land-holding class is constant. That implies a growth in net per-capita income of peasants. It is widely accepted that the *proportion* of total income devoted to demand or consumption of agrarian goods decreases as income grows. With total income consumed either as urban or agrarian goods the share consumed as urban goods, i.e. non-agrarian goods, increases as income grows. An increase in agrarian income will thus cause a changed composition of aggregate demand: a larger proportion of total income will be devoted to demand for urban goods. If we, furthermore, assume that labour mobility between sectors equalizes per-capita income levels such a change in the composition of demand will increase the urbanization ratio, the ratio of urban producers to the total number of producers. We abstract from sectoral differences in household size so that we can derive relative size relating to producers from relative size of population groups.

Unfortunately, we cannot observe changes in income and its composition directly but we have more reliable information concerning the urbanization ratio. And as can be recalled from the preceding paragraph the urbanization ratio tells us about the composition of consumption and hence the level of per-capita income. Although we proceed from information on urbanization to an estimate of income we need to introduce a theoretical concept to accomplish that transformation and we have to consider the distribution of income between peasants and landlords to relate income changes to productivity growth – but more about that essential step below. The theoretical concept needed to deduce an income increase from a rise in the urbanization ratio is the consumption function. The important, but uncontroversial, characteristic of the consumption function is that consumption of agrarian goods, more precisely the average propensity to consume agrarian goods, is a decreasing function of income:

$$c_t y_t = b + m y_t, \qquad (4.1)$$

where c is the average propensity to consume agrarian goods, y is net per-capita income in terms of agricultural product, b and m are time-invariant parameters; b is the intercept, m is the marginal propensity to consume agrarian goods and t is a time subscript. This consumption function implies that if a peasant household enjoys a higher net income than before then it will increase the consumption of agrarian and urban goods. But c, the average propensity to consume agrarian goods, will decline and the *proportion* of net income exchanged for urban goods, being $1-c$, will increase. The urbanization ratio, i.e. the ratio of urban population to total population, will increase as a consequence of the increased weight of urban goods in aggregate consumption.

How much a given increase in income will affect the urbanization ratio – or, conversely, the magnitude of the income increase inferred by an observed change in the urbanization ratio – depends on the marginal propensity to consume agrarian goods, m. The closer the marginal propensity is to its maximum, i.e. 1, the larger will the change in per-capita income be. The reason for this is straightforward: if almost all increases in income are spent on agrarian goods, then there must be a big absolute increase in income to accomplish a given increase in demand for urban products. And demand for urban goods is obviously the factor determining the relative proportion of urban producers in population, the urbanization ratio.[5]

But demand for urban products is also affected by changes in distribution of income. The distribution of income is important because the average propensity to consume agrarian goods differs between the land-holding classes and the urban and agrarian producers, since the land-holders enjoy much higher per-capita income and as a consequence a higher propensity to consume urban goods, i.e. $1-c$. If land-holders succeed in increasing the rate of rents the urbanization ratio will increase. In this case the rise in urbanization is not based on an increase in peasant income and labour productivity. By controlling for changes in distribution it will, however, be possible to establish an unambiguous relation between changes in urbanization and labour productivity.

To sum up: an observed change in the urbanization ratio is the basis for inferring changes in productivity once we have estimated a consumption function and controlled for changes in the distribution of income. The underlying logic of the argument is that an increase in net income decreases the average propensity to consume agrarian goods and increases the proportion of income spent on urban goods with an observed increase in urbanization as the result. An increase in the urbanization ratio that is caused by higher taxation will not be recorded as an increase in labour productivity because we control for changes in distribution. The magnitude of the productivity increase inferred by an observed change in the urbanization ratio will depend on the marginal propensity to consume agrarian goods and the distribution of income. (The way one can get from a net per-capita income measure to a gross, i.e. labour productivity, is discussed below.)

The consumption function is defined over a composite net product of agrarian goods, y, the net income, which makes it possible to avoid a series of familiar problems with productivity measurements. Since there is labour mobility between sectors, prices adjust and per-capita income is equal in the two sectors. But we need not take urban prices into account in estimating per-capita income since it is expressed in the net per-capita product of agrarian goods times the price of agrarian goods. The rationale behind this procedure is that we believe people in low-income countries first satisfy their basic need, food, and then use the residual for urban goods. The proportion devoted to agrarian consumption will, in other words, not be influenced by the prices on urban goods.[6] Furthermore it is not necessary to specify weights to each good produced. We do not even need to know the sort of products that have been cultivated. More generally, the problems of aggregation referred to above, cf. point 2, as well as the problem of including new goods, cf. point 3, or quality changes of known goods, cf. point 4, have been evaded. How can we be so sure about that? It all has to do with the consumption function that predicts a decline in the average propensity to consume agrarian goods if income grows. If, and only if, peasants experience an increase in income – irrespectively of whether the income growth arises from the introduction of a new and higher-valued good, better quality, more efficient storage of products (cf. point 5 above), or simply because more goods are being produced per labourer – they will react by devoting a larger share of their income to urban goods, evidence for which is a rise in urbanization. In a way, one can say that the problem of aggregation and estimation of income growth is solved by the economic agents themselves and will be reflected in the urbanization ratio.

Having now established an intuitive understanding of the method's logic, we can proceed to a detailed account with reference to the rigorous presentation in the appendix to this chapter. The first requirement is that the economy can be considered a closed unit in terms of agrarian production or, if it is not, that we have information about its agrarian trade balance. The economy under consideration has three classes and two producing sectors, an agrarian and an urban sector with agrarian and urban producers. Urban production is simply non-agrarian but need not necessarily be manufactured in cities. We will, however, use the observed urbanization ratio, i.e. urban to total population, as an approximation of the relative proportion of urban producers. In other words, it is assumed that the urban production in the countryside equals the agrarian production in cities and, as pointed out previously, that household sizes do not differ across sectors. There is an 'elite' composed of the land-holding classes inclusive of their servants, soldiers and officials. The land-holding classes receive rents from the agrarian producers and provide military protection (often a euphemism for oppression), political order and spiritual 'services'. The assumption that urban and agrarian per-capita income is equal is in no way essential for the argument. If we have good reasons to believe that, say, urban per-capita income was higher then that presumption is easily accommodated into the calculation (the effects on the results is discussed in section 4.3). Urban producers do not pay taxes to the land-holding classes but may or may not pay taxes to urban officials, who are part of the active urban population.[7]

The information we need for the calculation of labour productivity (i.e. agrarian product per labourer in the agrarian sector) is (1) the rate at which peasants are taxed (including rent and all sorts of levies in money, in kind or in labour); (2) the relative size of the three classes in society, i.e. peasants, urban producers and the land-holding classes (including the political and ecclesiastical elite, their servants, officials and soldiers) at the points in time between which we wish to make the productivity assessment; and finally (3) the share of total urban production absorbed by the land-holding classes at one point in time. We need not know either the actual *volume* of production, income, rent and taxes or the number of inhabitants.

112 Labour Productivity in the Mediaeval Economy

The actual procedure followed in making the calculations of labour productivity is briefly sketched here with references to the relevant parts of the appendix. The first step is to estimate the constant m in the consumption function, and an intermediary step in that process will be to derive s_t, the relative landlord per-capita income. That estimate depends on information at time t about the relative size of the agrarian producers and land-holding classes and of the rate of taxes and rents, r_t. The formula is derived in equation A4.6 in the appendix and states that s_t equals the ratio of rent to non-rent income multiplied by the ratio of peasant population to total land-holding population. Having derived s_0 for a specific point of time, say time 0 which is the initial year in the period for which we will make productivity calculations, we can estimate the constant m, i.e. the marginal propensity to consume agrarian goods. This is done by utilizing the following additional input of empirical data relating to time 0: the proportion of urban producers, which we know from the relative size of the two other classes, and secondly, the fraction k_0 of total urban production which is consumed by the land-holding classes. This additional information enables us to estimate m, the marginal propensity to consume agrarian goods (cf. equation A4.12 in the appendix).

With a known m we have an equation system of two equations (cf. equations A4.7 and A4.8 that imply A4.11 and A4.12) capable of generating values for the unknowns k_t (the fraction of aggregate urban product consumed by the land-holding classes) and b/y_t for any other date t for which it can be expected that the estimated consumption function applies. b/y_0 can also be estimated through equation A4.11 at time 0. b/y_t may at first sight seem to be an awkward expression, b is the intercept in the conception function, but if we have solved it for the initial year, time 0, and at the end, 1 of the period for which we are computing changes we get the rate of growth of the net income since

$$\frac{b/y_0}{b/y_1} = \frac{y_1}{y_0}. \tag{4.2}$$

Since net income is the share of per-capita output that is not expropriated by landlords, it can be expressed as $y_t = q_t(1-r_t)$ and labour productivity is thus $q_t = y_t/1 - r_t$. We get an expression for the change in labour productivity, $\hat{q}$,

$$\hat{q} = \frac{\dfrac{y_1}{1-r_1} - \dfrac{y_0}{1-r_0}}{\dfrac{y_0}{1-r_0}}, \tag{4.3}$$

which is what we have been trying to accomplish.

The advantages of the solution suggested here are that it provides us with a productivity concept that can be regarded as a welfare measure and that it is possible to avoid the problems associated with productivity measurement based on data on physical units. That achievement is not without its costs, however. The principal drawback of the method is that it must be assumed that the consumption function is stable over time, more precisely that the constants m and b do not change. At first sight this requirement may look very restrictive when we are investigating periods of several centuries. But the economy that we are discussing, although changing, is throughout the period a predominantly agrarian society in which the quest for food is a central concern. In an economy with low per-capita income the consumption function is determined by physiological and nutritional mechanisms rather than by psychological factors, and this will make it comparatively stable. We cannot of course neglect the possibility of shifts in the consumption function and we can predict the nature of such shifts if they occur. If m shifts at all it will plausibly become smaller since the increasing importance of cities and urban production may have demonstration and learning effects on men which gradually change preferences. That prediction is kept in mind when we interpret the results in section 4.3.

As is often the case when different approaches are evaluated, it is not only the intrinsic value that matters but rather the comparative advantage of an approach. It should be clear by now that the method presented here has many advantages in that it is free from the many technical problems associated with standard productivity measurements and makes it possible to study whole economies rather than only large estates. We proceed in the next section to present productivity estimates based on the available information. Given the present state of knowledge, the empirical information used for the calculations must be treated as approximations and the results as tentative. Needless to say, more robust results are obtained if the data base is improved. As is shown in section 4.3 the results that are obtained can be assessed by comparisons with results that are derived from alternative suggestions as to what the relevant data should be.

4.3 Some Tentative Results on the Evolution of Labour Productivity

The economy examined here is Western Europe, basically Italy, the Low Countries, Germany, France and England. It is an economy that is large enough to be considered a self-contained economic unit and it is culturally and economically homogenous enough to be regarded as a single unit. We will consider the period of great transformation from the end of the eleventh century, called time 0, to the beginning of the fourteenth century, called time 1. A broad outline of the technological changes, the growth of population and cities, division of labour and intensified trade has already been provided in chapters 1 and 3. The discussion in this chapter therefore concentrates on the more limited types of data needed for the calculations.

At the present state of knowledge one cannot do without a certain amount of *a priori* reasoning based, however, on rather uncontroversial principles. The range of uncertainty must be remembered when we come to interpret the results. These shortcomings are not, as was shown above, unique to the present method, which however yields results that are significant and meaningful.

At the start of the expansionary phase of the mediaeval economy there was an almost total dominance of agrarian producton. Cities were small and unimportant and what there was of urban production was mainly directed to the consumption of the land-holding classes. Land was still in abundant supply because of the low population density which – lacking precise information about rents in this early period – will be expected to lead to a comparatively low level of rents. Labour dues, which at this stage was the customary form of rent, of between one and two days a week were the rule. By the end of the thirteenth century the data are more abundant but, alas, not unambiguous. Undoubtedly, however, there was an increase in the urbanization ratio and rents tended to mount, which we would also expect from basic economic principles given the shortage of land.

TABLE 4.1 *Growth of labour productivity in Western Europe* (c.1100–c.1300)

Time 0 (c.1100)	
Empirical data	
Sectoral composition of population, proportions:	
Agrarian producers	0.9
Urban producers	0.05
Land-holding classes	0.05
The proportion of aggregate urban production consumed by the land-holding classes, k_0	0.8
The rate of rents, r_0	0.2
Data derived from the empirical information relating to time 0:	
Using A4.6 gives the relative landlord per-capita income, s_0	4.5
Using A4.11 and A4.12 generates the marginal propensity to consume agrarian goods, m	0.77
and b/y_0	0.22
Time 1 (c.1300)	
Empirical data	
Sectoral composition of the population, proportions:	
Agrarian producers	0.80
Urban producers	0.15
Land-holding classes	0.05
The rate of rents, r_1	0.35
Data derived from the empirical information relating to time 1:	
Using A4.6 gives the relative landlord per-capita income, s_1	8.6
and A4.11–14 generate	
b/y_1	0.167
y_1/y_0	1.32
k_1	0.56
$\hat{g}$	62%

Translating these qualitative arguments to figures suggests, as a first approximation, table 4.1.

The result, an increase of labour productivity of just over 60 per cent, suggests significant growth and is not based on controversially optimistic empirical estimates. Nonetheless, the plausibility of the empirical information used will be assessed and in some cases alternative empirical estimates will be used.

The general pattern that emerges from table 4.1 is one in which the urban production from having been almost exclusively oriented towards the land-holding classes increasingly becomes directed to satisfying demand from the agrarian and urban producers. Still at the end of the mediaeval expansion, time 1 in table 4.1, a little more than half of the urban production is luxury production, $k_1 = 0.56$. These findings seem to be plausible. The growth of net income, 30 per cent, is of a magnitude that will make the assumption of a stable consumption function less vulnerable to criticism. At very high increases in net income, say a doubling, one might expect a downward shift in m (the marginal propensity to consume agrarian goods) because of an increased exposure to new goods and habits (cf. above).

But is the marginal propensity used in table 4.1 reliable in the first place? The way we estimate m makes it dependent on the rate of taxation and rents and the relative size of the land-holding classes at time 0. The land-holding classes were, as pointed out above, composed not only of the elite itself but also included its servants, soldiers and officials. We do not have much information in this particular area, unfortunately. It is clear that the elite itself was very tiny, say between 0.5 and one per cent of the population which implies, assuming a proportion of 0.05 of the land-holding classes altogether, something like five to ten servants, soldiers and officials per member of the elite. The supporting members of the land-holding classes will most likely have a per-capita income equal to the net per-capita income of peasants, which, under the assumptions used for time 0, would give a member of the elite, i.e. the 0.5–1 per cent of the population, an income 15–25 times higher than enjoyed by peasants, apart from the services rendered by the supporting members. Income differentials of this magnitude are plausible which makes the estimates reasonable. What would the outcome be with a much lower m? If we take a marginal propensity to consume agrarian goods slightly above the one observed in some developing countries, say 0.6, the growth in labour productivity would be half of that in table 4.1, around 30 per cent. My conjecture is that this is at the lower end of a plausible interval of results.

With the customary estimate that the peasantry constituted 90 per cent of the population in the early phase of the mediaeval expansion, the land-holding classes cannot constitute much more than five per cent of the population. If the figure was somewhat higher, the recorded growth in productivity in table 4.1 would be an over-estimation. If, which is more likely, the land-holding classes are believed to be a smaller proportion, say 0.03, then the marginal propensity to consume agrarian goods estimated on the remaining information from table 4.1 would be somewhat higher (0.8), and the results in terms of net income and productivity a slight increase as well: net income will grow by almost 50 per cent and labour productivity by a little more than 80 per cent.

Some areas of Western Europe, big enough to be almost self-sufficient in agricultural products, attained higher urbanization ratios before the Black Death than the figure quoted in table 4.1. If we allow for a slightly higher urbanization ratio by the end of the period, say 0.18, the growth in productivity would be close to 100 per cent. But for these areas the marginal propensity constant in the consumption function may be inappropriately high. In the later phase of the mediaeval expansion and in areas with an urbanization ratio as high as the Low Countries (cf. table 3.2), the urban culture probably changed the preferences of the consumers. So if we consider Western Europe to have been a closed economy, the estimate of around 0.15 seems more reasonable.

If in the course of history the urban population enjoyed a relative rise (decline) in their per-capita income, the figures for productivity and income growth presented in table 4.1 will be an underestimate (overestimate).

There is some controversy around what may be the level of rents by the end of the thirteenth and early fourteenth century. Postan and followers suggest figures as high as 50 per cent, while van der Wee proposes a more modest figure of one-third of the gross agrarian grain yield (van der Wee, 1978, pp. 148–9). These assessments are very difficult to make because by the end of the thirteenth and the beginning of the fourteenth century the ownership pattern over land was very varied and therefore the incidence of taxation varied as well. While some peasants, or rather tenants, evidently paid up to 50 per cent, those less burdened paid only half of that (Titow, 1969, p. 81), and peasants with full property rights in their land may have been even less burdened. Peasants that entered crop-sharing arrangements were often exposed to unfavourable market conditions and can be expected to have experienced above-average exploitation. The evidence here suggests rents of between one-third and 50 per cent gross product (Dollinger, 1949, p. 135; Sivéry, 1973, pp. 375–81). Although the 'middle-of-the-road' estimate provided by van der Wee seems to be the most plausible estimate, a few calculations in what have been called the demographic approach and, alternatively, a Postan–Le Roy Ladurie scenario will be provided.

In this perspective, rents are believed to increase markedly because of sharply diminishing returns as population pressures mount. Furthermore, the urban regeneration is believed to be less important. Calculations based on these premises – here assumed as an increase in the urbanization ratio from 0.05 to 0.10 compared to 0.15 in table 4.1 and a significantly higher level of rents at time 1, $r_1 = 0.5$, i.e. at the end of the period – give some interesting results. Net income for peasants declines 30 per cent which seems implausible given the low level of net income at the start of the expansion. Such decline, furthermore, is without firm support in historical sources.

The general argument in the Postan–Le Roy Ladurie approach seems to be that incomes did not change much. Accepting this suggestion as a starting point we can derive the change in labour productivity associated with a sharp increase in rents but with constant peasant per-capita net income throughout the period by using equation A4.14. Setting $y_0 = y_1$ and $r_0 = 0.2$ and $r_1 = 0.5$ the necessary increase in labour productivity is 60 per cent, i.e. almost the same figure as generated by the data set from table 4.1. In this perspective, cities become *increasingly* geared towards luxury production, i.e. a k_1 close to 1 which does not seem to fit the available evidence of a growing market for non-luxury goods.

Contrary to the standard assumptions in the demographic approach, small changes in labour productivity, associated with a sharp decline in peasant net income provoked by augmenting rents, presuppose significant *technological* change. That is so because a positive, albeit small, increase in labour productivity is the compound effect of technological change and increased labour effort that is counterveiled by sharply diminishing returns in agriculture. The Postan–Le Roy Ladurie approach therefore seems neither theoretically consistent nor plausible in its empirical underpinnings.

On balance the calculations performed here suggest a significant increase in labour productivity in the expansion phase of the European mediaeval economy and presumably an important although less marked increase in peasant net income. This confirms that the model of self-sustained growth advocated in the preceding chapter has historical relevance.

Labour Productivity in the Mediaeval Economy   119

NOTES

1 As pointed out by G. Bois, mediaevalists have a perfect alibi in not discussing peasant production in a detailed fashion: 'Le paysan ne parle pas, ou plutôt, n'écrit pas' (Bois, 1976, p. 127).
2 See J. Goy's contribution to Le Roy Ladurie (1982) for a preservation of the project and the methodological problems involved in research of this kind. For national reports, see van der Wee (1978) and Goy (1982).
3 Recent studies include Titow (1972) and a theoretical and statistical critique in Desai (1977). See also Tits-Dieuaide (1975, especially ch. 2) and Derville (1987).
4 The evidence of cyclical changes in yields per unit of land is discussed in van der Wee and van der Cauwenberghe (1978, pp. 138–9).
5 E. A. Wrigley has applied a method similar to the one presented here for the estimation of labour productivity in early modern England, but without explicitly introducing the concept of a consumption function. His procedure is, however, based on an implicit assumption that the marginal propensity to consume agrarian goods is 0 and will therefore lead to an underestimation of productivity growth because m is certainly larger than 0. (See Wrigley 1967 and 1985.)
6 Let the price of agrarian goods be 1 and the price of urban goods, p. q_a is per-capita agrarian productivity and r the share of rents which makes the net per-capita income $y = (1-r)q_a$. Urban producers pay no taxes to landlords so the net per-capita income is equal to $q_u p$, and since prices adjust, the sectoral equality of income can be expressed as $y = q_u p$. If there is an increase in labour productivity in the urban sector, that will not affect y but only lead to a downward adjustment of urban prices, which amounts to saying that if urban producers spent all their output on agrarian goods they would get the same y as before the rise in productivity. This implies that y is not a real-income concept since it does not take into account that agrarian goods now 'buy' more urban goods and real income consequently has increased. So when we say that the consumption function is defined over y we mean that the share of y that is devoted to agrarian consumption is dependent on y only and not what urban goods y can buy. The advantage is of course that y can easily be transformed into a productivity measure, since (cf. above) $q_a = y/(1-r)$ and furthermore that y is not affected by changes in productivity in the urban sector.
7 See the appendix to this chapter for an explanation of why taxation in the urban sector does not affect the results.

APPENDIX

In this appendix the formulae used for calculation of changes in net income and labour productivity are derived. Intermediary steps in these calculations include the estimation of an index of relative landlord income, of the marginal propensity to consume agrarian goods, m, and of the proportion k of aggregate urban production consumed by the land-holding classes.

There are two producing sectors in the economy and three classes. Peasant households in the agrarian sector exchange part of their income for urban goods produced in the urban sector. Peasants pay rents to the land-holding classes who use it for their own consumption and for exchange with the urban sector.

Let q be the gross per-capita income in the agrarian sector and r the rate of taxes and rents. Then

$$y = q - rq, \tag{A4.1}$$

where y is net per-capita income. We also get a convenient expression for labour productivity, i.e. gross per-capita income in

$$q = \frac{y}{1-r}, \tag{A4.2}$$

$\hat{q}$ is the rate of growth of labour productivity, i.e.

$$\hat{q} = \Delta q/q. \tag{A4.3}$$

The consumption function has the familiar shape

$$cy = b + my, \tag{A4.4}$$

where c is the average propensity to consume agrarian goods and m and b are constants, m being the marginal propensity to consume agrarian goods. Multiplying b in A4.4 with $1-m/1-m$ provides a more useful expression for the consumption function

$$cy = \frac{b}{1-m} + m\left[y - \frac{b}{1-m}\right]. \tag{A4.5}$$

Let l_a be the proportion of the agrarian producers in the total active population, l_u the proportion of the urban producers and l_r the proportion of the landholding classes. The sum of $l_a + l_u + l_r = 1$ (in the actual calculations these proportions are assumed equal to the sectoral distribution of population, i.e. we abstract for differences in household size). More precisely $l_a = L_a/L$, $l_u = L_u/L$ and $l_r = L_r/L$ where L_a is the number of agrarian producers, L_u the number of urban producers, L_r the number of landholders and L finally the total number of producers. We can now proceed by defining an index s of per-capita landlord income expressed in terms of peasant per-capita net income, y.

$$s = \frac{rL_a\left[\dfrac{y}{1-r}\right]}{y} = \frac{r}{1-r}\frac{l_a}{l_r}. \tag{A4.6}$$

If we furthermore define k as the share of aggregate urban production consumed by the land-holding classes and assume that peasant net per-capita income equals urban per-capita income, then the following two equations hold

$$L_r\left[sy - \frac{b}{1-m}\right](1-m) = kL_u y, \tag{A4.7}$$

$$(L_a + L_u)\left[y - \frac{b}{1-m}\right](1-m) = (1-k)L_u y. \tag{A4.8}$$

Equation A4.7 states that landlords' consumption of urban goods is equal to their share of total urban production, and equation A4.8 that the consumption of urban goods by peasants and urban producers is equal to the part of total urban production not consumed by the land-holding classes. We can divide through by L in equations A4.7 and A4.8 so we can use the information on proportions rather than absolute numbers. We thus get

$$l_r\left[sy - \frac{b}{1-m}\right](1-m) = kl_u y, \tag{A4.9}$$

$$(l_a + l_u)\left[y - \frac{b}{1-m}\right] = (1-k)l_u y. \tag{A4.10}$$

Dividing equations A4.9 and A4.10 through with y, adding them and keeping one of the original equations provides

$$\frac{b}{y} = (1-m)(l_r s + l_a + l_u) - l_u, \tag{A4.11}$$

$$l_r s(1-m) - l_r \frac{b}{y} = kl_u. \tag{A4.10'}$$

A little bit of manipulation of A4.10' yields A4.12 in the equation system that will be used for the calculations

$$\frac{b}{y} = (1-m)(l_r s + l_a + l_u) - l_u, \quad (A4.11)$$

$$(1-m) = \frac{l_u}{s-1}\frac{1}{l_r}\left[k - \frac{l_r(1-k)}{l_a + l_u}\right]. \quad (A4.12)$$

We wish to get an estimate of labour productivity between two points in time, time 0 and 1. We do that by finding values for b/y_0 and $(1-m)$, the two unknowns at time 0. The known values of l_{a0}, l_{u0}, l_{r0}, k_0, and s_0, derived from A4.6, are substituted in A4.12 to get $(1-m)$. Then substitute $(1-m)$, l_{a0}, l_{u0}, l_{r0} and s_0 to find b/y_0 in A4.11.

At time 1 the unknowns are b/y_1 and k_1. We find b/y_1 by substituting the known values of $(1-m)$, l_{a1}, l_{u1}, l_{r1} and s_1 in A4.11. To find k_1 substitute $(1-m)$, l_{a1}, l_{u1}, l_{r1} and s_1 in A4.12.

With both b/y_0 and b/y_1 we can estimate the growth of net income since

$$\frac{y_1}{y_0} = \frac{\frac{b}{y_0}}{\frac{b}{y_1}}. \quad (A4.13)$$

Recalling that $q = y/1 - r$ we can get the change in labour productivity, $\hat{q}$, between time 0 and 1 through

$$\hat{q} = \frac{\frac{y_1}{1-r_1} - \frac{y_0}{1-r_0}}{\frac{y_0}{1-r_0}}, \quad (A4.14)$$

which was what we wished to accomplish.

If we admit for taxation in the urban sector, that will not affect the result. Assume that

$$L_u = L_p + L_b, \quad (A4.15)$$

where L_u is total active urban population, L_p is producers of urban goods and L_b is a group of bureaucrats supported by taxes from the urban producers. Net per-capita income is y, i.e. equal to agrarian net income. Let y_p be gross per-capita income of producers of urban goods.

Then
$$U = L_p y_p, \tag{A4.16}$$
where U is aggregate urban income but y is net average urban income,
$$y = \frac{U}{L_u}, \tag{A4.17}$$
which implies that
$$U = L_u y, \tag{A4.18}$$
which is what is stated in equations A4.7 and A4.8.

5 Why have Growth Rates been so Low until Recently?

5.1 Introduction

The preceding chapters have presented theoretical arguments and empirical support for the view that technological progress is a systematic force in human civilization. This view of ongoing technological change raises a number of related questions some of which are answered here and some of which are put on the agenda for further inquiry.

If technological progress is pervasive in human history why has it been so feeble during most of the past history? This problem will be addressed and it will be argued that the slowness of technological progress is compatible with the present framework. Section 5.2 presents an account of the determinants of the rate of technological growth. The role of labour and the social relations under which labour is performed are emphasized in section 5.3. Section 5.4 takes up the question of periodic stagnation and set-back. Technological progress is the rule in economic history but interruptions and temporary reversals of this process do occur and they can be accounted for within the present approach. In section 5.5, finally, it is argued that far from being a theory of low growth only the present framework is applicable well into the industrial revolution. That invites us to seek an explanation of changes in the rate of growth.

5.2 The Determinants of Technological Progress

In chapter 1, pre-industrial technological change was said to be based on growth of knowledge which is endogenous in production and dependent on (1) economies of practice that, to a certain extent at least, are transferable from one producer to another and from one generation to another. The economies of practice gained are directly related to the number of times a certain productive task is performed per period and producer. That condition also applies to (2) trial-and-error experiments, and (3) stochastic 'mutations' of known methods. Some of these save resources and are subsequently put into practice by producers. Finally, (4) technological change was shown to be related to specialization and division of labour, the division of labour being positively related to the aggregate demand, to 'the extent of the market' as Adam Smith put it. This relation arises because of technological characteristics of both production and education: there are indivisibilities in equipment and in learning superior skills and these fixed costs cannot be covered unless demand permits production at considerable level.

Let us pause a moment to consider the nature of the division of labour process. Woodwork is a good example because it highlights the parallel differentiation of crafts and tools. Like most other technologies woodwork had suffered a set-back in the post-Roman period but signs of recovery are visible by the eleventh century. At that time, however, all woodwork was done by carpenters who were in charge of both the building of houses and the making of furniture. During the mediaeval and early modern period an ongoing specialization took place. Manufacturing of furniture and housebuilding developed into separate crafts and within each there emerged more specialized occupational groups such as joiners, turners and carvers (within furniture making) and within housebuilding the role of carpenters was circumscribed by workers in charge of the making of panels (which were important as a device for insulation). This craft differentiation required specialized tools for the particular tasks performed and some tools came in different functional designs to match the specific demands of each skill – see Symonds (1956) for a survey of mediaeval woodwork.

Division of labour therefore implies, firstly, a division of the production process into an increasing number of separable tasks and, secondly, a concentration by each producer on fewer of the separable tasks. It will consequently be accompanied by an increase in the total number of identical productive tasks per producer and period, an increase in what will be called the compound specialization level.[1]

Economies of practice, trial and error and stochastic 'mutations' are all positively related to the compound specialization level. The reasons for this were discussed in chapter 1 and can be briefly recapitulated here. When it comes to economies of practice the reason is straightforward. As suggested by the terminology, the gains arise because of the very repetitive nature and the degree of repetition is enhanced by division of labour: a producer performs fewer tasks but each task is done more often. Turning to the problem of stochastic effects there are clearly both non-beneficial and economizing ones. A full variance of beneficial 'disturbances' will occur only when the number of times a given type of productive operation (task) is performed in a production unit per period is high. The larger the number of times the faster will new useful knowledge be gained. This point is important because if productive operations were performed non-frequently the beneficial 'mutations' might be forgotten and the producer might be provided with a less rich menu of beneficial 'mutations'. Finally, the incentives for trial-and-error experiments will increase if the specific type of task is performed often. The costs of trials are that they sometimes fail to produce an improvement and may, in fact, diminish the expected output. But since the costs of such failures can be measured in the labour effort forgone in a trial, the cost will be relatively small as producers concentrate their time on fewer tasks done repeatedly. Furthermore, any positive pay-off from trial-and-error experiments may weigh more heavily if that particular task is performed very often. As in the case of 'mutations' the producer gets a richer menu of trials to choose from.

The rate of endogenous technological progress will be positively related to the level of per-capita income and the size of the economy (i.e. the number of members in it) because the compound specialization level is. More precisely, these factors affect both the growth of knowledge and its diffusion. By definition, the aggregate production is determined by the size of the economy and the per-capita output; let us therefore look more closely into the way these factors are correlated with technological progress. A rise in per-capita income will not only directly affect aggregate production (given the size of the population) but also increase the rate of population growth. The consumption pattern will furthermore be transformed by growth in per-capita income leaving a larger proportion of income for exchange which in turn influences the extent of the market and generates an increase in demand for non-agrarian goods. There are also some vital differences between non-agrarian and agrarian production processes that make the latter less susceptible to division of labour. No doubt there is great potential for regional specialization within agriculture taking advantage of natural, climatological and geographical conditions (i.e. availability of markets and natural means of transport, such as vicinity to rivers and sea). Improved conditions for exchange will furthermore stimulate a concentration in peasant communities on agrarian tasks only, leaving the non-agrarian production, for example some processing of food or tool-making, to specialists. But there are obvious constraints on division of labour as well. The production process is to a large extent dictated by nature while industrial activities are only restrained by human ingenuity. We would therefore expect non-agrarian production to be more malleable by division of labour leading to an almost infinite specialization of production into separable tasks. Nonetheless, agrarian technological progress has been impressive during periods with rapid transformation of the industrial trades.[2]

The growth of the size of an economy evades simplified accounts. It is not only a matter of demographic factors but political and technological ones as well. Low levels of technology necessarily impose limits on the size of the economy because the extensive land use generates low population density. There is also the problem of establishing a social order so that a common technological and cultural heritage is effectively shared and diffused and transmitted from one generation to the other. Apart from the technological and political determinants of the size of the population there are purely demographic forces. There is a link between growth of per-capita income and growth of population, but sometimes, and especially so in pre-industrial economies, the causality is disrupted by exogenous forces such as epidemics and disorder. As is briefly discussed below, such decline in the size and order of societies usually affects technological progress negatively.

To sum up: the rate of growth of new knowledge from economies of practice, trial and error and chance variation in established methods will be related to the level of per-capita income and the size of the economy, i.e. the aggregate income. But the stock of knowledge only represents a potential for technological progress while realized technological progress is determined by how much and how fast the stock of knowledge is put into practice, i.e. the rate of diffusion.

What exactly is the significance of the rate of diffusion? It bridges the gap between potential and realized technological progress. If all new knowledge was applied without delay wherever it was applicable we would not have any problem with diffusion. But not all existing knowledge is used by all and what is used is applied only after a delay.

An increase in the rate of diffusion speeds up the process in which new knowledge is applied and increases the fraction of knowledge that is put to use. In short, an increased rate of diffusion narrows the gap between best and average practice.

When the new experience is recorded in isolated and small production units realized technological progress will necessarily be unimpressive. Only when producers become members of larger economic networks – which presupposes exchange and hence relatively high per-capita income – they also become exposed to a richer flow of experience and knowledge. And as they experience increasing per-capita income, producers will exchange a greater proportion of their income with other producers. The rate of diffusion is also linked with the social relations of production. It is important that the direct producers, who discern and record the new knowledge, have incentives to remember it as well as power and opportunities to implement it.

Can we go from this general framework to an indicator, albeit approximative, of the rate of technological progress? There is in fact one emerging from the arguments in the previous paragraphs. The important link between the economic factors such as aggregate production and per-capita product and technological progress is the level of compound specialization. It would be useful if we could observe that level. This is unfortunately a difficult empirical task to accomplish. But it was suggested that the compound specialization effect is the result of two distinct processes that are inversely related to each other (cf. note 1) and amounts to a concentration of producers on fewer tasks each of which is performed more frequently. So in fact it would be sufficient to know the number of separable tasks, for which the division of labour may serve as an indicator. If we observed an increase in the division of labour or an increase in the number of occupational groups we would be tempted to suggest that the rate of technological progress had increased as well. (But some doubts will be raised against a too self-confident interpretation).

5.3 Labour as a Productive Force

So far labour has figured only as an agent in the specialization process. That does not imply, however, that labour is reduced to a passive role in economic development. The purpose of this section is not to discuss the importance of individualism when – for reasons discussed extensively in chapter 2 – it became socially accepted as the size of the economies grew. The arguments advanced in chapter 2 still stand and they do strengthen the case for our stress on the size of the economy as a decisive factor. Instead I shall focus on the social relations under which labour is performed as vital for the rate of diffusion of knowledge. If workers are alienated from the control of the production process and denied freedom of contract and exchange, as is the case with slavery, and to a large extent serfdom, then several consequences follow.

The experience which is generated endogenously in production through stochastic 'mutations' will be weakly absorbed by producers since they do not have any incentives to remember and apply it. A possible exception would be 'mutations' that saved labour effort – but not necessarily labour time since the latter was often fixed. For similar reasons, trial and error would not be practised on an extensive scale. And the gains that are made in terms of relevant knowledge will only slowly be applied because the direct producers have little say in these matters. When those possessing the most elaborate knowledge of the production process are unable to communicate with the world outside the *latifundia* or the manor, the diffusion process is curbed even more. To some extent these defects of slavery and serfdom were compensated for by managers with intimate knowledge of the production process. But the supervisory structure was very costly and much information can be expected to have been lost because serfs did not feel committed to the same objectives as the managers. Demanorialization is therefore an important change in social relations because the direct producers are put at the centre of the stage: those who receive new knowledge are free to exchange it and will be rewarded when putting it into practice.

5.4 The Historical Significance of Piecemeal Technological Progress

The explanation put forward here of technological progress is simple but it is neither trivial nor tautological. While it is true by definition that an economy that initially starts at a low per-capita income level and experiences very low technological progress will remain rather poor, it is not true by definition that low incomes explain low rates of technological progress. A specific causal relationship has been introduced: it has been argued that the growth of knowledge is related to per-capita and aggregate income. Nor does the argument amount to the trivial observation that division of labour increases labour productivity.

Although both true and important this connection between division of labour and the level of productivity will not in itself affect the permanent rate of technological progress. The hypothesis advanced in this book is, on the other hand, that the *rate* of technological progress, not just the level, is related to per-capita and aggregate income. This relation is based on the influence of division of labour on stochastic 'mutations', and trial and error, as well as on the speed of learning by doing and to what extent (and speed) new knowledge is diffused. More importantly, it would then be possible to explain the low rate of technological progress without resort to *ad hoc* explanations which impose peculiar mentalities on man, e.g. an inherent conservatism, or, as is sometimes implied in the property-rights approach, a particular inability of pre-industrial man to create a proper institutional setting for production and exchange. This is not to say that on a more general level the problem of social institutions (rather than the property rights in a more restricted sense) are unimportant. On the contrary the institutional setting is decisive; it does, for instance, regulate the size of the economy and thereby directly influences aggregate production and division of labour.

Mankind has spent most of history in hunter-gatherer technology and the question must be posed why the transition to agriculture took such a long time. There are several plausible explanations that are compatible with the growth-of-knowledge approach defended in this book. Hunter societies are for obvious technological reasons very small and contacts with other groups are infrequent and in many cases belligerent. The archaeological record does not permit us to say anything about when the sort of institutions that are adequate for cultural continuity emerged. In the absence of exclusive territorial rights, however, societies of hunters and gatherers will eventually face the problem of unrestrained competition for scarce resources, conflict and the disintegration of a technological heritage. We cannot tell whether institutional failures prolonged the hunter–gatherer era. More importantly, the very size and isolation of these societies suggest an extremely slow rate of technological progress, although such progress should not be ruled out entirely. As discussed in chapter 1, the hunter technology is a parasitic one. The population depends on the available biomass and technological change is primarily of a leisure-increasing variety. Only to the extent that an increasing proportion of the total biomass is exploited is an increase in population density possible.

Neither of these two types of technological progress should be dismissed. They both represent an increasing understanding of the environment which ultimately makes the transition to agriculture possible.

Any society at a low level of technology is, however, susceptible to exogenous shocks, epidemics and climatological changes. From this follows as a general conjecture that the very long pre-neolithic era might depend on the difficulties which these societies encountered in creating long stretches of cultural continuity necessary to accumulate an advanced technological heritage.

The story of the far-reaching consequences of the agrarian transition is a familiar one and there is no need to tell it again. Agriculture enhances population growth, division of labour and specialization. Even if it took several thousands of years of agricultural practice before the first advanced cultures appeared, it is important to stress that there was progress in many areas and that this progress permitted a comparatively steady growth in population. When population growth was combined with political institutions that increased the territorial size and stability of the economy we can also observe signs of change in the level of productivity that may be interpreted as a change in the rate of technological progress. But other interpretations are also possible. It may be the effect of an unchanged rate of progress combined with longer stretches of cultural continuity making for an increasing productivity or one-off productivity increases related to increased division of labour.

By modern standards the advances made in the Roman era and in the Chinese dynasties and earlier agrarian cultures in west Asia are slow but when comparison is made with the previous millenia of agrarian and metallurgic technology the change is impressive. The most obvious accomplishment of the Roman culture – which borrowed freely from the previous Mediterranean cultures – is its refined urbanism. There were great improvements and innovations in construction technology, in house-building and in water supply as well as a high level of division of labour. In Waltzing's effort to identify the different occupations in Rome something like 150 urban crafts are specified (Carcopino, 1986, pp. 195–202). That is a degree of specialization which is similar to that found in large mediaeval cities. If we use the number of separable tasks as an indicator of the speed of technological progress, a procedure proposed above, then one would expect signs of technological progress.

And there certainly was considerable development in viticulture and agriculture (e.g. improved efficiency in implements such as wine and olive presses, spades, ploughs and mechanical harvesters, traction power, irrigation, conservation of food, especially wine that kept up to 15 years). Many of these improvements or innovations were based on the increased use of iron (for example, in spades necessary for drainage work), and this use of iron relied on considerable progress in mining and metallurgy. The intensive land use demanded a hitherto unseen use of natural manure and chemical fertilizers (nitrate). In townplanning, leisure and hygiene (the baths and theatres) the progress is almost too obvious to mention (Duval, 1962, pp. 218–54). Recalling the arguments pursued in chapters 3 and 4 it is evident that the level of urbanization, which was not surpassed until mediaeval times, must also have been associated with a prolonged period of rise in per-capita income and hence a positive rate of technological progress (although not necessarily an increasing rate, of course).

The Roman accomplishment has often been overshadowed by what it did not achieve, e.g. the slow diffusion of the water-mill. But was that necessarily a sign of a technological inertia? C. Parain has suggested a series of explanations as to why the mill technology was not widely used in Europe until the dynamic phase of the mediaeval economy. Since a mill constituted a considerable indivisibility, the type of regional centralization provided by the feudal organization was better suited for it than the *latifundia*. Given the fact that it was not widely diffused in the Roman period the subsequent decline in population may have turned the indivisibility into a technological barrier. Its introduction in Europe on a wide scale was accompanied by the rise in population. Furthermore, there was a modification of corn types and a change in consumption patterns in that corn products were used increasingly for bread rather than porridge. Both these changes required proper milling rather than crushing by means of a piston. Finally, a change in income distribution in favour of the masses may have influenced these transformations (Parain, 1979, pp. 311–18).

The restricted use of the water-mill in the Roman era has been contrasted with the wide application it found in China at the same time. The water-mill seems to appear at approximately the same time (shortly BC) in Europe and China and this simultaneity supports the view of independent evolution. The hypothesis of independence is strengthened by the different uses of the water-mill. In China, the water-mill was not primarily used for agricultural purposes but in metallurgy and later in textile manufacturing. Its early introduction into metallurgy contributed to the rapid development of that technological sequence in China – again in contrast to the decline in post-Roman Europe. Cast iron and steel were manufactured in the first centuries AD when water-mill-driven bellows made it possible to control the pyrotechnical processes necessary for the sophisticated metallurgy, so that by the sixth century AD something akin to a Siemens–Martin process was applied. These early advancements were probably based on the ability to convert rotary motion to longitudinal (Needham, 1969, p. 20). The mastering of water power continued to distinguish the Chinese achievement and by the beginning of the early modern period China was technologically more advanced than Europe in almost any area, although at that time Europe had begun to catch up. Several centuries before similar attempts in Europe, the Chinese textile industry was mechanized using elaborate methods of conversion of motion, e.g. the crank.

The background for these divergent developments is possible to explain within this framework. Let us first look at population growth. While fertility rates were largely responsive to economic well-being, it was shown in chapter 3 that mortality rates were often very sensitive to exogenous shocks, such as epidemic cycles. It is believed that epidemics reduced the total population and aggregate demand considerably for several centuries after the demise of the Roman Empire. This reduction implies a de-specialization which has a one-off effect on labour productivity and per-capita income and possibly a slow-down effect on the rate of technological progress (assuming that the rate of technological progress is positively related to the level of per-capita income and the size of the economy).

What we witness complies well with these sombre expectations. There were no significant technological innovations in Europe until the end of the first millenium. In fact there are several signs of retrenchment. With the decline of urban population and reduced demand, specialization decreased. There was also a technological contraction in mining and as a consequence a diminished use of iron in farm implements. Viticulture suffered serious set-backs as did some labour-intensive farming methods such as manuring and marling. Some of the technological innovations of the Roman period were rediscovered in the mediaeval period, e.g. the iron-reinforced spade. Other Roman innovations were refined when they were put into practice again; examples include viticulture, drainage, marling and manuring in agriculture, irrigation, construction techniques and road-building, navigation and land transport.

There has long since been a debate among historians about whether the political disintegration of the Roman Empire should be seen as endogenous, i.e. as caused by some sort of institutional failure, such as overtaxation or labour supply problems rather than merely by exogenous aggression (cf., for example, Cipolla, 1970). This is an important issue but it is not one which can be settled here. What matters in the present context is that, whatever the reasons, political disintegration is bound to reduce the size of the economy and thereby provoke a (further) decline in aggregate demand, in specialization and ultimately in the level of income and productivity.

The reason why China developed along what J. Needham, the foremost historian of Chinese technology, has called 'a relatively slowly rising curve' while Europe stagnated, is straightforward. China experienced a remarkable stability, the essential prerequisite for cultural continuity, and more favourable demographic conditions with several periods of expansion. Estimated population reached the remarkable figure of 100 million by 1100 AD. It also appears that producers, at least those engaged in urban activities, enjoyed considerable freedom and status and produced for a mass market (Needham, 1969).

So far, we have primarily discussed stagnation as caused by exogenous shocks such as exogenously induced decline in population and political disorder. It must be pointed out, however, that this is not the whole story. At a low level of technological sophistication, the diminishing returns associated with a fixed supply of land may be felt at comparatively low densities of population. The analysis in chapter 3 showed that diminishing returns will be countervailed by technological progress but that economies may experience stagnating per-capita income if technological progress is feeble. Such Malthusian equilibria may however be characterized by incomes above subsistence level and associated with population growth. The risk of stagnating per-capita income (or declining per-capita income if the diminishing returns are aggravated by population growth) may thus be greater at low levels of income, implying lower rates of technological progress, than at more advanced stages. The present framework can therefore accommodate the historical fact that economies have experienced low and constant per-capita income for long stretches of history. But as has been repeatedly stressed stagnating per-capita income associated with diminishing returns presupposes technological progress.

Summing up the arguments, the present framework can reconcile the idea that technological progress is a pervasive force in human civilization with the fact that it has been slow until the most recent centuries. Furthermore, as we shall see in the final section, the framework is also capable of explaining the gradual increase in growth rates of per-capita income.

5.5 Continuity and Revolution

The framework for the analysis of technological progress as endogenous in production suggests a continuity in the growth of knowledge, at a rate affected by the level of per-capita income and the size of the economy. How does this stress on continuity square with the frequent assertions of revolutions in the economic history of mankind? We have already referred to the neolithic revolution, the transition towards agriculture, and in addition there have been claims of an industrial revolution in the thirteenth century (Carus–Wilson, 1941), an agrarian revolution in the sixteenth and early seventeenth centuries (Kerridge, 1967) instead of that which was normally believed to have occurred in the eighteenth century. And of course there is the industrial revolution of the eighteenth century, which Rostow called the take-off into sustained growth.

The long-term consequences of the agrarian transition have been far-reaching since it permitted a continuous growth of population, but as argued in chapter 1, the transition itself was not a sudden and swift transformation. Likewise, the increased importance of textile and other types of manufacturing was remarkable in thirteenth-century Europe and affected industrial employment, technology and productivity. But it still seems to be – from a technological point of view – a refinement and some new adaptations of known implements and, to some extent, the productive use of new knowledge gained in production. So continuity there was. What made these transformations revolutionary was not a new pattern of technological progress but only more favourable conditions for accumulation and diffusion of knowledge: long periods of cultural continuity and demographic growth without severe exogenous disturbances in terms of aggressions or epidemics. It is more controversial, perhaps, to claim continuity throughout the eighteenth-century British economy, in the century of the industrial revolution. But in fact it can be – and has been – argued that it is not until the middle of the nineteenth century that a new and potent force makes its entry and changes the nature of technological progress radically. The force in question is the direct relationship between science and production. Until the middle of the last century science and purposeful search for new technology based on scientific discovery was of minor importance, if not insignificant. Students of technological history, such as A. Hall, D. Landes and L. Mumford have long since been aware of this, and N. Rosenberg (with L. E. Birdzell) expressed the point succinctly when they wrote:

> ...at the beginning of the nineteenth century, most industrial technology, including the Industrial Revolution, was the work of artisans and engineers with little or no scientific training. Shipbuilding, engineering, construction, architecture, mining, smelting, weaving, and the other industrial arts of 1800 were based on experience, rules of thumb, and craft tradition. (1986, p. 244)

But if technological progress was based on a slowly growing body of knowledge gained through experience rather than theoretical inquiry, how can we explain the marked increase in productivity ascribed to the industrial revolution? The answer is that there was no such abrupt increase in productivity. Although there has been a growing awareness among economic historians of the inflated credentials given to the industrial revolution, a concept so well-entrenched will not be easily abandoned. The concept of the industrial revolution was a vivid characterization of a period which was meant to distinguish it from the alleged inertia of the pre-industrial epochs.

In the traditional chronicle, the main impact of the revolutionary transformations released by industrialism occurred in the second half of the eighteenth century. The changes involved both manufacturing and agriculture and marked a dividing line between old and new patterns of growth. A series of recent reconsiderations of the statistical basis for this general conclusion generate a very different picture, however. There is no breakthrough to a significantly higher growth rate by the second half of the eighteenth century. In fact there is a fairly steady growth throughout the eighteenth century of around 0.3–0.4 per cent annual growth of GDP per capita. The rate increases to 0.5 per cent in the first third of the nineteenth century. Although these figures represent a dramatic downward revision they still may be slightly over-estimated for reasons discussed in the note accompanying table 5.1 below. The story of the industrial revolution as a sudden transition to high growth is an outcome of a familiar index problem. The spectacular and new industries (cotton, steam and steel) were in fact relatively unimportant in terms of relative shares but were given far too high weights in the estimates provided by Cole and Deane (1962). They grew much faster than the traditional occupations and therefore distorted the whole picture. When the fast-growing industries were given more appropriate weights the growth process that emerged was a less turbulent one (Harley, 1982; Crafts, 1983, 1985; Lee, 1986, especially ch. 1). Furthermore, the agrarian nature of early eighteenth-century England was over-estimated. By 1700, the non-agrarian population in England may have constituted close to 40 per cent, which is much higher than earlier believed. The implication of these revisions is that the industrialization or proto-industrialization must have started in the preceding centuries (Lindert, 1980).

If we turn our attention to agriculture, there is also good reason to stress the continuity in a sequence of steady progress. Kerridge (1967) has challenged the conventional wisdom by arguing that the great transformation of English agriculture occurred in the sixteenth and seventeenth centuries. He shows that English agriculture adopted some of the practices already used on the continent, such as convertible husbandry (i.e. permanently cultivated arable alternating between temporary tillage and temporary grass leys), floating of meadows which gives early grass, systematic use of a wider variety of manure and fertilizers, fen drainage and better selection of (new) crops and livestock. Kerridge may have exaggerated the impact of some of the recorded transformations (Outhwaite, 1986) since some new crops and practices were introduced rather late in his chosen period. There is no doubt however that the view of a transformation of English agriculture as concentrated in the period of the alleged take-off into industrial revolution must be discarded. It is equally doubtful that the recorded innovations were imported from the Low Countries which admittedly were more advanced at this time (as in the mediaeval period). It is not necessary to apply the diffusionist interpretation. Most of the new crops grown were native to Britain or known long since, and practices such as floating meadows and convertible husbandry 'owed nothing to foreign practices' (Clay, 1984, vol. I, p. 130). Furthermore, as pointed out in chapter 3, some of the crops and rotation systems were already used in the mediaeval era in certain areas of England and the Continent. The nature of the technological improvements was of the endogenous and piecemeal type and may therefore appear in many different locations independently. The nature of the process has been well described by an historian of the remarkable Dutch accomplishment in the sixteenth and seventeenth centuries but the argument applies elsewhere as well:

> Breeding improvements before the eighteenth century occurred rather fortuitously. The absence of accurate scientific knowledge about this matter did not prevent farmers from experimenting. (de Vries, 1974, p. 142)

There are notable advances in other areas such as crop strains with better resistance and yields. There is probably a more rapid diffusion due to the increased commercialization and market opportunities, and the urge for perfection may have been enhanced, but these gradual changes are typically the outcome of an endogenous growth-of-knowledge process.

The general pattern that emerges in both agriculture and manufacturing is one of continuity (interrupted by population decline and wars) rather than sudden and abrupt changes. In table 5.1 an attempt is made to estimate the growth of per-capita income (or productivity) over a very long period. The pattern that emerges is very much akin to the one suggested here from theoretical considerations: there is a low but rising rate of growth and if there is a 'leap forward' it is in the early modern period rather than in the eighteenth century. This might reflect the increasing sophistication of the division of labour, the growth in size of the economies through internationalization and trade, with implications for the scope of individualistic pursuits and diffusion of knowledge.

Needless to say, the computation of annual figures over very long periods does not imply that the growth rate was constant throughout the period. We know, in fact, that it was not. The period in the aftermath of the Black Death has not even been included because of the general uncertainty about how to interpret what happened. There are some signs of increasing per-capita income but uncertainty as to what happened to urbanization. A constant urbanization ratio in the face of

TABLE 5.1 *Annual growth of per-capita income in England (1100–1830)* (%)

(1) 1100–1300	(2) 1500–1700	(3) 1700–60	(4) 1760–80	(5) 1780–1801	(6) 1801–31
0.1–0.24	0.28–0.34	0.31	0.01	0.35	0.52

Sources and notes: The income (= product) concept is not entirely consistent throughout the period. In columns 1 and 2 the per-capita income is expressed in agrarian gross per-capita product while columns 3–6 show national product per capita. The income in columns 1 and 2 is not a real-income concept and will reveal an underestimate (over-estimate) if productivity growth in the urban sector was faster (slower) than in the agrarian sector. Column 1 suggests that the plausible result is within a range where the higher value is based on table 4.1 and on the presumption that the stylized facts concerning Western Europe in that table apply to England as well. The estimated growth of gross per-capita income was 62 per cent and a dynamic phase of 200 years generated the recorded figure. The lower estimate in column 1 is based on the assumption of lower marginal propensity to consume agrarian goods, m = 0.6, and a period of self-sustaining growth of 250 years. Column 2 is based on a method described in the appendix to this chapter. It is similar to the one used for the calculation of table 4.1 with two exceptions. There are only two classes: agrarian and urban and the agrarian class entails landlords. That implies that we do not have to take the distribution of income between peasants and landlords into account. Furthermore, we do not estimate the marginal propensity to consume agrarian goods but use data from present-day developing countries on a similar level of development. The average propensity to consume agrarian goods (food) is used as an indicator of the level of development. The share of the agrarian population in the total population can be interpreted as an approximation of the average propensity to consume agrarian goods, cf. the appendix to this chapter for an elaboration of this point. The agrarian population was estimated to be 80 per cent in 1500 and 60 per cent in 1700. The lower estimate in column 2 is generated by giving the agrarian population in 1500 a relative size of 75 per cent. Crafts (1985, pp. 11–17, table 2.1) suggests in his appraisal of the Lindert–Williamson (1982) reinterpretation of the social tables of 1688 an even lower share of the agrarian population, close to 55 per cent. The higher figure of 60 per cent is used because the material is based on households, and one would suspect a higher participation of women in agriculture than in non-agrarian activities. Developing countries with an average propensity to consume agrarian goods around 60 per cent included Korea, Thailand and the Phillipines (in the 1950s and 1960s) and their marginal propensities ranged from 48 to 31 per cent, while countries such as Italy with a lower average propensity (46 per cent) had a marginal propensity around 40 per cent (Lluch, 1977, tables 3.3 and 3.6). The marginal propensity used for the calculations of column 2 is 0.4. Finally, columns 3–6 are from Crafts (1985, table 2.11). See Persson (1988) for an appraisal of Crafts' calculations.

increasing per-capita income may, however, be explained by a growing income differential in favour of urban dwellers. The basis for that conjecture can be that mobility could not match the higher mortality in cities in periods of epidemics (Persson, 1984). The results must be seen as tentative or, using a phrase borrowed from Crafts' characterization of his calculations, 'controlled conjectures'. To the extent that a critique raises doubts as to the empirical validity of the information used, the method used for columns 1 and 2 can easily accommodate other sets of data.

The pattern which the table suggests, is one of a slow upward trend in growth. The rates of growth are within a range that is consistent with a view of technological progress as the outcome of an endogenous although intensified growth of knowledge.[3] Although the figures presented are 'controlled conjectures' their plausibility is strengthened if they can be confirmed by theoretical expectations. The underlying determinants of change are trial and error, learning by doing, and stochastic 'mutations' stimulated by division of labour. These factors are in turn enhanced by the very size of an increasingly international economy permitting greater individual freedom (see chapter 2). Production and demand is directed into areas of production more malleable to division of labour and areas more susceptible to human ingenuity and to competitive pressures. There is reason to believe that growth and diffusion and application of knowledge will be enhanced, but not to the extent that a break with past history is generated. The reason for that expectation is simply that there is not as yet a radical change in the determinants of technological progress which occurs with the forceful intervention of science in technology in the second half of the nineteenth century. That is why the downward revision of the inflated growth figures associated with the traditional accounts of the industrial revolution make sense. Instead of looking at the industrial transformation by the end of the eighteenth century as the beginning of a new era it can better be understood as the culmination of technology's *ancien régime*. At last economic history and the history of technology are telling the same story.

NOTES

1 Although this argument should be intuitively clear a little mathematics can help clarify the matter. Let Q be aggregate production and L the number of producers, and q is per-capita product (gross income), then

(i) $q = Q/L$.

The number of separable tasks per unit of output is t so that T is the total number of tasks performed

(ii) $T = Qt$.

Division of labour implies that t is increasing but also that the number of separable tasks each producer performs diminishes, what we can call a concentration effect. Call the number of separable tasks a producer performs c and we obtain an expression for the compound specialization level, i.e. the number of times a producer performs an identical task

(iii) $s = (Qt/L)/c$.

But since $Q = Lq$ (cf. (i)), one can write (iii) as

(iv) $s = qt/c$.

If q is increasing and there is Malthusian population growth, Q will also grow (i.e. if we disregard a falling L because of exogenous influences such as epidemics) and assuming that an increasing proportion of q is exchanged, we can easily see that the extent of the market will grow. The implication will be that t is increasing and c is decreasing. s will consequently grow.

2 Although the analysis suggested here has an obvious classical flavour the economists of that tradition were quite pessimistic about the long-run prospects of the industrial economy. To a large extent that pessimism was due to their belief that diminishing returns ruled agrarian production and would finally inhibit the growth of the entire economy. In that belief they were already wrong in their own lifetime. See Wrigley (1987, ch. 2) on the classical economists and the industrial revolution.
3 Although the economy portrayed in table 5.1 has the stylized characteristics of the English economy it shares the basic trends with other advanced parts of Europe until the late seventeenth century or early eighteenth century.

APPENDIX

The economy has two sectors, an agrarian and a non-agrarian, the latter composed of manufacturing, transport, services and commerce. There is balanced trade in agrarian goods. The factual accuracy of the latter proposition is not easily established. In the seventeenth century there was a trade deficit in foodstuffs but a huge surplus in woollens which had a large agrarian content, so the assumption is not far off the mark. The gross per-capita income q is the same in both sectors and the consumption function is defined over the gross per-capita product measured in agrarian goods. Note, however, that the effects of an increasing (decreasing) urban/rural income differential is discussed in the footnote to table 5.1. The consumption function now reads

$$c_t q_t = b + m q_t, \tag{A5.1}$$

in which c is the average propensity to consume agrarian goods, b is a positive constant and m the marginal propensity to consume agrarian goods. The corresponding consumption function for time 1 is

$$c_1 q_1 = b + m q_1, \quad (A5.2)$$

but it is convenient to express q_1 by the following identity

$$q_1 = q_0 + (q_1 - q_0). \quad (A5.3)$$

If A5.3 is inserted in A5.2 we get

$$c_1 q_1 = b + m q_0 + m(q_1 - q_0), \quad (A5.4)$$

which can be simplified using A5.1 to

$$c_1 q_1 = c_0 q_0 + m(q_1 - q_0). \quad (A5.5)$$

Equation A5.5 is used to derive the results in column 2 of table 5.1 by making q_1 the unknown and setting q_0 at 100. The growth in productivity expressed as a percentage is given by $(q_1 - q_0)/q_0 \times 100$. m is empirically given from economies on a comparative level of average propensity to consume agrarian goods (cf. the discussion in chapter 5), and c_t is derived from the relative distribution of the population between the two sectors. The average propensity to consume agrarian goods, c, is equal to the share of the agrarian population in the total population. Let L be total population and L_a the agrarian population. Then the following identity holds

$$c_t L_t q_t = L_{at} q_t, \quad (A5.6)$$

stating that the total consumption of agrarian goods in the economy is equal to the production of agrarian goods. But from A5.5 follows by simplification that

$$c_t = L_{at}/L_t, \quad (A5.7)$$

which makes it possible to estimate c from information about the relative share of the agrarian population.

References and Bibliography

Abel, W. (1967) *Geschichte der Deutschen Landwirtschaft*. Stuttgart: Eugen Ulmer.
Abel, W. (1980a) *Agricultural Fluctuations in Europe from the Thirteenth to the Twentieth Centuries*. London: Methuen.
Abel, W. (1980b) Deutsche Agrarwirtschaft im Hochmittelalter, in Kellenbenz (1980), pp. 539–47.
Aerts, E. and van der Wee, H. (1982) *De economische ontwikkeling van Europa 950–1950*. Leuven: Acco.
Alchian, A. A. (1950) Uncertainty, evolution and economic theory. *Journal of Political Economy*, pp. 211–21.
Ammerman, A. J. and Cavalli-Sforza, L. L. (1971) Measuring the rate of spread of early farming in Europe. *Man*, pp. 674–88.
Arrow, K. (1962) The economic implications of learning by doing. *Review of Economic Studies*, pp. 155–73.
Bairoch, P. (1985) *De Jéricho à Mexico. Villes et économie dans l'histoire*. Paris: Gallimard.
Baldwin, J. W. (1959) The medieval theories of the just price. Romanists, canonists and theologians in the twelfth and thirteenth century. *Transactions of the American Philosophical Society*, vol. 49, part 4. Philadelphia: The American Philosophical Society.
Barker, G. (1985) *Prehistoric Farming in Europe*. Cambridge: Cambridge University Press.
Barker, G. and Gamble, C. (1985) *Beyond Domestication in Prehistoric Europe*. London: Academic Press.
Bender, B. (1975) *Farming in Prehistory: From Hunter–gatherer to Food-producer* London: John Barber.
Binford, L. R. (1968) Post-Pleistocene adaptions, in Binford and Binford (1968).
Binford, L. R. and Binford, S. R. (1968) *New Perspectives in Archeology*. Chicago: Aldine.
Bois, G. (1976) *Crise du féodalisme*. Paris: Presses de la Fondation Nationale des Sciences Politiques.
Bois, G. (1978) Against the neo-Malthusian orthodoxy. *Past and Present*, pp. 60–9.
Boserup, E. (1965) *The Conditions of Agricultural Growth: The Conditions of Agrarian Change under Population Pressure*. London: Allen and Unwin.
Boserup, E. (1981) *Population and Technology*. Oxford: Basil Blackwell.

Boutruche, R. (1970) *Seigneurie et féodalité. L'apogée XIe–XIIIe siècle*. Paris: Aubier.
Braidwood, L. and Braidwood, R. (1969) Current thoughts on the beginnings of food production in southwestern Asia. *Mélanges de l'Université Saint Joseph*.
Braidwood, R. J. and Howe, B. (1960) *Prehistoric Investigations in Iraqui Kurdistan*. Studies in Ancient Oriental Civilization, vol. 31. Chicago: University of Chicago Press.
Brants, V. (1895) *L'économie politique au Moyen Age: ésquisses des théories économique professés par les écrivains des 13e et 14e siècles*. Louvain: C. Peeters.
Braverman, A. and Stiglitz, J. E. (1986) Landlords, tenants and technological innovations. *Journal of Development Economics*, pp. 313–32.
Brenner, R. (1976) Agrarian class structure and economic development in pre-industrial Europe. *Past and Present*, pp. 30–75.
Brenner, R. (1982) The agrarian roots of European capitalism. *Past and Present*, pp. 16–113.
Britnell, R. H. (1981) The proliferation of markets in England 1200–1349. *Economic History Review*, pp. 209–21.
Byock, J. L. (1982) *Feud in the Icelandic Saga*. Berkeley, Calif.: University of California Press.
Campbell, B. M. S. (1983a) Agrarian productivity in medieval England: some evidence from Norfolk. *Journal of Economic History*, pp. 379–403.
Campbell, B. M. S. (1983b) Agricultural progress in medieval England: some evidence from eastern Norfolk. *Economic History Review*, pp. 26–45.
Campbell, B. M. S. (1984) Population pressure, inheritance and the land market in a fourteenth-century peasant community, in Smith (1984).
Campbell, B. M. S. (1987) Arable productivity in medieval English agriculture. Unpublished paper.
Carcopino, J. (1986) *Daily Life in Ancient Rome. The People and the City at the Height of the Empire*. Harmondsworth: Penguin (first published in English in 1941).
Carus-Wilson, E. M. (1941) An industrial revolution of the thirteenth century. *Economic History Review*, pp. 39–60.
Cashdan, E. A. (1985) Coping with risk: reciprocity among the Basarawa of northern Botswana. *Man*, pp. 454–74.
Charles, J. A. (1980) The coming of copper and copper base alloys and iron: a metallurgic sequence, in Wertime and Muhly (1980), pp. 151–82.
Childe, G. V. (1952) *New Light on the Most Ancient East*. London: Routledge and Kegan Paul.
Chittolini, G. and Coppola, G. (1982) Grand domaine et petites exploitations: quelques observations sur la version italienne de ce modèle, in Gunst (1982).
Cipolla, C. (1947) Comment s'est perdue la propriété écclésiastique dans l'Italie du Nord entre le XIe et le XVIe siècle. *Annales Economies, Sociétés, Civilisations*, pp. 317–27.
Cipolla, C. (1949) The trends in Italian economic history in the late Middle Ages. *Economic History Review*, pp. 181–4.
Cipolla, C. (ed.) (1970) *The Economic Decline of Empires*. London: Methuen.
Cipolla, C. (ed.) (1972) *The Fontana Economic History of Europe: The Middle Ages*. London: Fontana.

Cipolla, C. (ed.) (1974) *The Fontana Economic History of Europe: The 16th and the 17th Centuries.* London: Fontana.
Clark, G. and Pigott, S. (1970) *Prehistoric Societies.* Harmondsworth: Penguin (first published 1965).
Clay, C. G. A. (1984) *Economic Expansion and Social Change: England 1500–1700*, vol. 1: *People, Land and Towns*, vol. 2: *Industry, Trade and Government.* Cambridge: Cambridge University Press.
Coase, R. H. (1960) The problem of social cost. *Journal of Law and Economics*, pp. 1–44.
Cohen, G. A. (1978). *Karl Marx's Theory of History: A Defence.* Oxford: Oxford University Press.
Cohen, G. A. (1980) Functional explanation: reply to Elster. *Political Studies*, pp. 129–35.
Cohen, M. N. (1977) *The Food Crisis in Prehistory, Overpopulation and the Origins of Agriculture.* New Haven: Yale University Press.
Cole, W. A. and Dean, P. (1962) *British Economic Growth 1688–1959, Trends and Structure.* Cambridge: Cambridge University Press.
Crafts, N. F. R. (1983) British economic growth, 1700–1831: a review of the evidence. *Economic History Review*, pp. 177–99.
Crafts, N. F. R. (1985) *British Economic Growth During the Industrial Revolution.* Oxford: Oxford University Press.
Daelemans, F. (1982) Tithe revenues in rural south-western Brabant, fifteenth to eighteenth centuries, in van der Wee (1978).
Dahlman, C. J. (1980) *The Open Field System and Beyond: A Property Rights Analysis of an Economic Institution.* Cambridge: Cambridge University Press.
Daumas, M. (ed.) (1962) *Histoire générale des techniques* vol. 1: *Les origines de la civilisation technique.* Paris: Presses Universitaires de France.
Deane, P. and Cole, W. A. (1967) *British Economic Growth 1688–1959.* Cambridge: Cambridge University Press.
Demsetz, H. (1967) Towards a theory of property rights. *American Economic Review*, pp. 347–59.
de Roover, R. (1958) The concept of the just price: theory and economic policy. *Journal of Economic History*, pp. 418–39.
de Roover, R. (1967) *San Bernardino of Siena and Sant' Antonio of Florence. Two Great Economic Thinkers of the Middle Ages.* Kress Library of Business and Economics. Publication no. 19. Boston.
Derville, A. (1978) La réduction des jachères au Moyen Age dans la Flandre Wallone. *Bulletin du Centre d'Etudes Médiévales et Dialectales de l'Université de Lille*, vol. 3, no. 1: *Bien dire et bien apprendre.*
Derville, A. (1982) Seigneur Jehan Florent de la Porte, mayeur de Saint-Omer, et ses exploitations rurales, *Revue du Nord*, pp. 683–700.
Derville, A. (1987) Dîmes, rendements du blé et révolution agricole dans le nord de la France au Moyen Age. *Annales Economies, Sociétés, Civilisations*, pp. 1411–32.
Desai, M. (1977) *A Malthusian Crisis in Medieval England: A Critique of the Postan–Titow Hypothesis.* London School of Economics (mimeograph).
Desportes, P. (1977) *Reims et les reimois aux XIIIème et XIVème siècles*, two vols. Reims: Service de Reproduction des Thèses, Université de Lille.

de Vries, J. (1974) *The Dutch Economy in the Golden Age 1500–1700*. New Haven: Yale University Press.
Dobb, M. (1946) *Studies in the Development of Capitalism*. London: Routledge and Kegan Paul.
Dodghson, R. (1981) The interpretation of subdivided fields: a study in private and communal interests?, in Rowley (1981).
Dollinger, P. (1949) *L'évolution des classes rurales en Bavarie. Depuis la fin de l'époque carolingienne jusqu'au milieu du XIIIe siècle*. Paris: Publications de la Faculté des Lettres de l'Université de Strasbourg.
Domar, E. (1970) The causes of slavery or serfdom: a hypothesis. *Journal of Economic History*, pp. 18–32.
Dowd, D. F. (1961) The economic expansion of Lombardy 1300–1500. *Journal of Economic History*, pp. 143–60.
Duby, G. (1962) *L'économie rurale et la vie des campagnes dans l'occident médiévale*. Paris: Aubier.
Duval, P. M. (1962) L'apport technique des romains, in Daumas (1962), pp. 218–54.
Dyer, C. (1980) *Lords and Peasants in a Changing Society. The Estates of the Bishopric of Worcester*. Cambridge: Cambridge University Press.
Elster, J. (1980) Cohen on Marx's theory of history. *Political Studies*, pp. 121–8.
Elster, J. (1983) *Explaining Technical Change*. Cambridge and Oslo: Cambridge University Press and Universitetsforlaget.
Fenoaltea, S. (1976) Risk, transaction costs, and the organization of medieval agriculture. *Explorations in Economic History*, pp. 129–51.
Fenoaltea, S. (1977) Fenoaltea on open fields: a reply. *Explorations in Economic History*, pp. 405–10.
Fourquin, G. (1972) *Le paysan d'occident au Moyen Age*. Paris: Fernand Nathan.
Furubotn, E. G. and Pejovich, S. (1972) Property rights and economic theory: a survey of recent literature. *Journal of Economic Literature*, pp. 1137–62.
Fussel, G. E. (1966) Ploughs and ploughing before 1800. *Agricultural History*, pp. 177–86.
Glass, D. V. and Revelle, R. (eds) (1972) *Population and Social Change*. London: Edward Arnold.
Gordon, B. (1975) *Economic Analysis before Adam Smith*. London: Macmillan.
Goy, J. and Le Roy Ladurie, E. (eds) (1982) *Prestations paysannes, dîmes, rente foncière et mouvement de la production agricole à l'époque préindustrielle*. Paris: Mouton.
Gross, C. (1890) *The Guild Merchant. A Contribution to British Municipal History*, vol. 1. Oxford: Clarendon Press.
Gunst, P. and Hoffmann, T. (eds) (1982) *Large Estates and Small Holdings in Europe in the Middle Ages and Modern Times*. Budapest: Akademiai Kiado.
Habakkuk, H. J. (1958) The economic history of modern Britain. *Journal of Economic History*, pp. 486–501.
Halfpenny, P. (1981) Two-variable and three-variable functional explanations. *Philosophy of the Social Sciences*, pp. 27–32.
Hahn, F. (1982) Reflections on the invisible hand. *Lloyds Bank Review*, pp. 1–21.

Hallam, H. E. (1981) *Rural England 1066–1348*. Glasgow: Fontana.
Harley, C. K. (1982) British industrialization before 1841: evidence of slower growth during the industrial revolution. *Journal of Economic History*, pp. 267–89.
Harvey, B. F. (1966) The population trend in England between 1300 and 1348. *Transactions of the Royal Historical Society*, 5th series, pp. 23–42.
Hatcher, J. (1977) *Plague, Population and the English Economy 1348–1530*. London: Macmillan.
Hatcher, J. (1981) English serfdom and villeinage: towards a reassessment. *Past and Present*, pp. 3–39.
Heers, J. (1973) *L'Occident aux XIVe et XVe siècles. Aspects économiques et sociaux*. Paris: Presses Universitaires de France.
Hicks, J. (1969) *A Theory of Economic History*. Oxford: Oxford University Press.
Hilton, R. H. (1969) *The Decline of Serfdom in Medieval England*. London: Macmillan.
Hilton, R. H. (ed.) (1976) *Transition from Feudalism to Capitalism*. London: New Left Books.
Hilton, R. (1982) Towns in societies. *Urban History Yearbook*, pp. 7–13.
Imberciadori, I. (1980) Italien: Die Landwirtschaft von 11. bis 14. Jahrhundert, in Kellenbenz (1980), pp. 441–50.
Irsigler, F. (1982) Die Gestaltung der Kulturlandschaft am Niederrhein unter dem Einfluss städtischer Wirtschaft, in H. Kellenbenz (ed.) *Wirtschaftsentwicklung und Umweltbeeinflussung, Beiträge zur Wirtschafts- und Sozialgeschichte*, Band 20. Weisbaden: Steiner.
Irsigler, F. (1983) Die Auflösung der Villifikationsverfassung und der Übergang zum Zeitpachtssystem im Nahbereich niederrheinischer Städte während des 13./14. Jahrhunderts, in H. Patze (ed.) *Die Grundherrschaft im späten Mittelalter, Vorträge und Forschungen*, Band 27. Sigmaringen: Jan Thorbecke Verlag.
Jope, E. M. (1956) Agricultural implements, in Singer et al. (1956).
Keene, D. (1985) *A Survey of Documentary Sources for Property Holding in London before the Great Fire*. London: London Record Society.
Kellenbenz, H. (ed.) (1980) *Handbuch der Europäischen Wirtschafts- und Sozialgeschichte*, Band 2. Stuttgart: Klett-Cotta.
Kerridge, E. (1967) *The Agricultural Revolution*. London: Allen and Unwin.
King, E. (1973) *Peterborough Abbey 1086–1310: A Study in the Land Market*. Cambridge: Cambridge University Press.
Klemm, F. (1959) *A History of Western Technology*. London: Allen and Unwin.
Kussmaul, A. (1981) *Servants in Husbandry in Early Modern England*. Cambridge: Cambridge University Press.
Langdon, J. (1982) The economics of horses and oxen in medieval England. *Agricultural History Review*, pp. 31–40.
Langdon, J. (1986) *Horses, Oxen and Technological Innovation: The use of Draught Animals in English Farming from 1066 to 1500*. Cambridge: Cambridge University Press.
Laslett, P. (ed.) (1972) *Household and Family in Past Time*. Cambridge: Cambridge University Press.
Lee, C. H. (1986) *The British Economy since 1700: A Macroeconomic Perspective*. Cambridge: Cambridge University Press.

Le Goff, J. (1986) *La bourse et la vie: Economie et religion au Moyen Age*. Paris: Hachette.
Le Roy Ladurie, E. (1966) *Les paysans du Languedoc*. Paris: Mouton.
Le Roy Ladurie, E. and Goy, J. (1982) *Tithe and Agrarian History from the Fourteenth to the Nineteenth Centuries: An Essay in Comparative History*. Cambridge: Cambridge University Press.
Lindert, P. H. (1980) English occupations, 1670–1811. *Journal of Economic History*, pp. 685–712.
Lindert, P. H. and Williamson, J. G. (1982) Revising England's social tables 1688–1712. *Explorations in Economic History*, pp. 385–408.
Lluch, C., Powell, A. A. and Williams, R. A. (1977) *Patterns in Household Demand and Saving*. Washington: Oxford University Press/The World Bank.
McCloskey, D. (1976) English open fields as behaviour towards risk. *Research in Economic History*, pp. 124–70.
McCloskey, D. (1977) Fenoaltea on open fields: a comment. *Explorations in Economic History*, pp. 402–4.
McCloskey, D. and Nash, J. (1984) Corn at interest: the extent and cost of grain storage in medieval England. *American Economic Review*, pp. 174–87.
Macfarlane, A. (1978) *The Origins of English Individualism: The Family, Property and Social Transition*. Oxford: Basil Blackwell.
Mane, P. (1983) *Calendriers et techniques agricoles. France–Italie, XIIe–XIIIe siècles*. Paris: Le Sycomore.
Mathias, P. (1983) *The First Industrial Nation: An Economic History of Britain*, 2nd edn. London: Methuen.
Mauss, M. (1954) *The Gift: Forms and Functions of Exchange in Archaic Societies*. London: Cohen & West.
Mazzaoui, M. F. (1981) *The Italian Cotton Industry in the Later Middle Ages*. Cambridge: Cambridge University Press.
Miani, G. (1964) L'Economie lombarde aux XIVe et XVe siècles. Une exception à la règle? *Annales, Economie, Sociétés, Civilisation*, pp. 569–79.
Mickwitz, G. (1936) *Die Kartellfunktionen der Zünfte und ihre Bedeutung bei der Entstehung des Zunftwesens. Eine Studie in spätantiker und mittelalterlicher Wirtschaftsgeschichte*. Helsinki and Leipzig: Societas Scientarium Fennica.
Miller, E. and Hatcher, J. (1978) *Medieval England: Rural Society and Economic Change 1086–1348*. London: Longman.
Miller, W. I. (1986) Gift, sale, payment, raid: case studies in the negotiation and classification of exchange in medieval Iceland. *Speculum*, pp. 18–50.
Mols, R. (1982) Population in Europe 1500–1700: two centuries of demographic evolution, in Cipolla (1974).
Monroe, A. E. (1924) *Early Economic Thought*. Cambridge, Mass: Harvard University Press.
Morineau, M. (1977) Cambrésis et Hainaut: des frères ennemis? *Revue Historique*, pp. 323–43.
Myrdal, J. (1986) *Medeltidens åkerbruk, agrarteknik i Sverige ca 1000 till 1520*. Nordiska Muséets Handlingar, vol. 105. Stockholm: Nordiska Muséet.

Nagel, E. (1977) Functional explanations in biology. *Journal of Philosophy*, pp. 280–301.
Needham, J. (1969) *The Grand Titration: Science and Society in East and West*. London: Allen and Unwin.
Needham, J. (1970) *Clerks and Craftsmen in China and the West*. Cambridge: Cambridge University Press.
Needham, J. (1980) The evolution of iron and steel technology in east and southeast Asia, in Wertime and Muhly (1980), pp. 507–41.
Nelson, R. R. and Winter, S. (1982) *An Evolutionary Theory of Economic Change*. Cambridge, Mass.: Harvard University Press.
North, D. C. (1981) *Structure and Change in Economic History*. New York: W. W. Norton.
North, D. C. and Thomas, R. P. (1973) *The Rise of the Western World: A New Economic History*. Cambridge: Cambridge University Press.
O'Brien, G. (1920) *An Essay on Medieval Economic Thinking*. New York: Longmans, Green and Co.
Outhwaite, R. B. (1986) Progress and backwardness in English agriculture, 1500–1650. *Economic History Review*, 2nd series, pp. 1–18.
Parain, C. (1966) The evolution of agricultural technique. In Postan (1966), ch. 2.
Parain, C. (1979) *Outils, ethnies et développement historique*. Paris: Editions Sociales.
Persson, K. G. (1984) Labour and leisure in the late Middle Ages, in D. Menjot (ed.) *Manger et boire au Moyen Age*. Nice: Les Belles Lettres.
Persson, K. G. (1988) *Aggregate Output and Labour Productivity in English Agriculture 1688–1801*. University of Copenhagen (mimeograph).
Phillips, P. (1981). *The Prehistory of Europe*. Harmondsworth: Penguin (first published 1980).
Pirenne, H. (1925) *Medieval Cities, their Origins and the Revival of Trade*. Princeton: Princeton University Press.
Pirenne, H. (1936) *Economic and Social History of Medieval Europe*. London: Kegan Paul, Trench, Trubner.
Polanyi, K. (1944) *The Great Transformation*. New York: Farrar and Reinhart.
Poos, L. R. (1985) The rural population of Essex in the later Middle Ages. *Economic History Review*, pp. 515–30.
Postan, M. M. (ed.) (1966) *Cambridge Economic History of Europe*, vol. 1, 2nd edn. Cambridge: Cambridge University Press. Including M. M. Postan Medieval agrarian society in its prime: England.
Postan, M. M. (1972) *The Medieval Economy and Society: A History of Britain in the Middle Ages*. London: Weidenfeld and Nicholson.
Postan, M. M. (1973) *Essays on Medieval Agriculture and General Problems of the Medieval Economy*. Cambridge: Cambridge University Press.
Radner, R. (1986) The internal economy of large firms. *Economic Journal* (Conference Papers), pp. 1–22.
Razi, Z. (1980) *Life, Marriage and Death in a Medieval Parish*. Cambridge: Cambridge University Press.
Renfrew, C. (1969) The autonomy of the south-east European copper age, *Proceedings of the Prehistoric Society*, pp. 12–47.

Renfrew, C. (1976) *Before Civilization: The Radiocarbon Revolution and Prehistoric Europe*. Harmondsworth: Penguin (first published in 1973).
Rosenberg, N. (1982) *Inside the Black Box: Technology and Economics*. Cambridge: Cambridge University Press.
Rosenberg, N. and Birdzell, L. E. (1986) *How the West Grew Rich: The Economic Transformation of the Industrial World*. London: I. B. Tauris.
Rotelli, C. (1973) *Una campagna medievale. Storia agraria del Piemonte fra il 1250 e el 1450*. Turin: G. Einaudi.
Rowley, T. (ed.) (1981) *The Origins of the Open-Field Agriculture*. London: Croom Helm.
Russel, J. C. (1972) Population in Europe 500–1500, in Cipolla (1972).
Sahlins, M. (1974) *Stone Age Economics*. London: Tavistock Publications.
Schneider, H. (1979) *Livestock and Equality in East Africa*. Bloomington.: Indiana University Press.
Schotter, A. and Schwödiauer, G. (1980) Economics and the theory of games: a survey. *Journal of Economic Literature*, pp. 479–527.
Seebohm, F. (1896) *The English Village Community Examined in its Relation to the Manorial and Tribal Systems and to the Common or Open Field System of Husbandry: An Essay in Economic History*. London: Longmans, Green and Co.
Semenov, S. A. (1964) *Prehistoric Technology: An Experimental Study of the Oldest Tools and Artefacts from Traces of Manufacture and Wear*. London: Cory, Adams and Mackay.
Service, E. (1966) *The Hunters*. Englewood Cliffs, N.J.: Prentice Hall.
Singer, C., Holmyard, E. J. and Hall, A. R. (eds) (1956) *A History of Technology*, vol. 2. Oxford: Oxford University Press.
Sivéry, G. (1973) *Structure agraire et vie rurale dans le Hainaut à la fin du Moyen Age*. Lille: Service de Reproduction de Thèses, Université de Lille.
Sivéry, G. (1976) Les profits de l'éleveur et du cultivateur dans le Hainaut à la fin du Moyen Age. *Annales, Economies, Sociétés, Civilisations*, pp. 604–30.
Sivéry, G. (1982) Le début de l'économie cyclique dans les bassins Scaldien et Mosan, fin du XIIe et début du XIIIe siècle. *Revue du Nord*, pp. 667–81.
Sivéry, G. (1984) *L'économie du royaume au siècle de Saint-Louis*. Lille: Presses Universitaires de Lille.
Slicher van Bath, B. (1963) *An Agrarian History of Europe AD 500–1850*. London: Edward Arnold.
Slicher van Bath, B. (1972) Historical Demography and the Social and Economic Development of the Netherlands, in Glass and Revelle (1972).
Smith, A. (1776) *The Wealth of Nations*. Harmondsworth: Penguin (1982).
Smith, R. M. (1983) Hypothèses sur la nuptialité en Angleterre aux XIIIe–XIVe siècles. *Annales, Economies, Sociétés, Civilisations*, pp. 107–36.
Smith, R. M. (ed.) (1984) *Land, Kinship and Life-Cycle*. Cambridge: Cambridge University Press.
Spooner, B. (1972) *Population Growth: Anthropological Implications*. Cambridge, Mass.: MIT Press.
Struever, S. (1971) *Prehistoric Agriculture*. Garden City, N.Y.: The Natural History Press.

Sweezy, P. (1950) The transition from feudalism to capitalism, *Science & Society*, pp. 134–57, also in Hilton (1976), pp. 33–56.
Symonds, R. W. (1956) Furniture: post-Roman, in Singer et al. (1956), pp. 240–58.
Tawney, R. H. (1926) *Religion and the Rise of Capitalism: A Historical Study*. London: John Murray.
Thirsk, J. (1987) *England's Agricultural Regions and Agrarian History, 1500–1750*. London: Macmillan.
Thrupp, S. (1963) Economic organization and policies in the Middle Ages, in M. M. Postan and E. E. Rich (eds) (1963) *Cambridge Economic History of Europe*. vol. 3. Cambridge: Cambridge University Press.
Thrupp, S. (1972) Medieval industry 1000–1500, in Cipolla (1972).
Titow, J. Z. (1969) *English Rural Society 1200–1350*. London: Allen and Unwin.
Titow, J. Z. (1972) *Winchester Yields: A Study in Medieval Agricultural Productivity*. Cambridge: Cambridge University Press.
Tits-Dieuaide, M.-J. (1975) *La formation des prix céréaliers en Brabant et en Flandre au XVe siècle*. Bruxelles: Editions de l'Université de Bruxelles.
Tylecote, R. F. (1980) Furnaces, crucibles, and slags, in Wertime and Muhly (1980), pp. 183–228.
Unwin, G. (1908) *The Guilds and Companies of London*. London: Methuen.
van der Wee, H. and van der Cauwenberghe, E. (eds) (1978) *Productivity of Land and Agricultural Innovation in the Low Countries 1250–1800*. Leuven: Belgish Centrum voor Landelijke Geschiedenis.
van der Woude, A. M. (1982a) Population developments in the northern Netherlands. *Annales de Demographie Historique*, pp. 55–72.
van der Woude, A. M. (1982b) Large estates and small holdings: lords and peasants in the Netherlands during the late Middle Ages and early modern times, in Gunst (1982), pp. 193–207.
van Houtte, J. A. (1977) *An Economic History of the Low Countries 800–1800*. London: Weidenfeld and Nicholson.
van Houtte, J. A. (1980) Europäische Wirtschaft und Gesellschaft von den grossen Wanderungen bis zum Schwarzen Tod, in Kellenbenz (1980), pp. 1–149.
van Parijs, P. (1981) *Evolutionary Explanation in the Social Sciences, An Emerging Paradigm*. Ottawa: Rowman and Littlefield.
Verhulst, A. (1963) L'économie rurale de la Flandre et la dépression économique du Bas Moyen Age. *Etudes Rurales*, pp. 68–80.
Verhulst, A. (1985) L'intensification et la commercialisation de l'agriculture dans les Pays-Bas méridionaux au XIIIe siècle, in *La Belgique rurale du Moyen-Age à nos jours. Mélanges offerts à Jean-Jacques Hoebaux*. Bruxelles: Editions de l'Université de Bruxelles.
Vinogradoff, P. (1892) *Villeinage in England: Essays in English Medieval History*. Oxford: Clarendon Press.
Weber, M. (1958) *The Protestant Ethic and the Spirit of Capitalism*. New York: Charles Scribner and Sons.
Wertime, T. A. and Muhly, J. D. (1980) *The Coming of the Age of Iron*. New Haven: Yale University Press.
White, Lynn Jr (1962) *Medieval Technology and Social Change*. Oxford: Clarendon Press.

White, Lynn Jr (1978) *Medieval Religion and Technology: Collected Essays.* Berkeley, Calif.: University of California Press.
Whittle, A. (1985) *Neolithic Europe: A Survey.* Cambridge: Cambridge University Press.
Wolff, P. (1986) *Automne du Moyen Age ou printemps des temps nouveaux: L'économie européenne aux XIVe et XVe siècles.* Paris: Aubier.
Wrigley, E. A. (1967) A simple model of London's importance in changing English society and economy, 1650–1750. *Past and Present* pp. 44–70. Also in Wrigley (1987).
Wrigley, E. A. (1985) Urban growth and agricultural change: England and the Continent in the early modern period. *Journal of Interdisciplinary History,* pp. 683–728. Also in Wrigley (1987).
Wrigley, E. A. (1987) *People, Cities and Wealth: The Transformation of Traditional Society.* Oxford: Basil Blackwell.
Wrigley, E. A. and Schofield, R. S. (1981) *The Population History of England 1541–1871. A Reconstruction.* London: Edward Arnold.
Yelling, J. A. (1982) Rationality in the common fields. *Economic History Review,* pp. 409–15.

Index

Aage, H., viii
Africa
   metallurgy, 23
   open-field agriculture, 45
aggregate demand, 28, 34
   decline in, 134
   and division of labour, 10–11, 31
   and specialization, 70
   and technological progress, 88, 125–7
aggregate income, 129–30
   and technological progress, 125–7
aggregation in productivity measurements, 106, 110–11
agrarian goods, consumption of, 109, 112, 116
agriculture
   climatico-ecological view of the transition to, 15
   continuity of development, 137
   density of sowing, 105
   distribution in feudal society, 73
   gardening, 79, 81
   horses, 28–30, 77, 85
   innovations (before the Black Death), 79, 81
   irrigation, 79
   land-use efficiency, 26–7
   property relations, 67–70
   quality of plants, 25, 31, 138
   regional specialization, 31
   rotation and fallowing regimes, 14, 24, 27, 30, 65, 78, 81, 85, 106, 138
   transition to, 13, 15, 19, 21, 28

Althing, 43
altruism, 42, 60
Ammerman, A. J., 15
anti-Malthusian views, 12, 72
apprenticeships, 9
Arrow, K., 9
Asia, west, 15, 20–1, 28
Atkinson, T., viii
bargaining
   characteristics of, 36, 43
   in small economies, 40, 43, 51, 54
   and transaction costs, 51
Barker, G., 20
best practice, 8–9, 69
   and average practice, 70, 128
Birdzell, L. E., 4, 9
Black Death, 73, 76–7, 79, 82–3, 102, 117, 138
Bois, G., 63, 89
Boserup, E., 16–17
Boutruche, R., 83
Braidwood, R. J., 14
Brenner, R., 63–4, 73
British Isles, agrarian transition, 21

Campbell, B. M. S., vii, 82
capitalism, 64, 68
Cavalli-Sforza, L. L., 15
Childe, V. G., 15
China
   metallurgy, 21, 23, 133
   technological progress and cultural continuity, 131–4
   watermills, 132

Church, 51–2
Cipolla, C., 78
Clark, G., 14
Coase, R. H., 34
Cobb–Douglas function, 91
Cohen, G. A., vii, 5, 37
Cohen, M. N., 16
commercialization approach, 64, 99
  and specialization, 66
common property, over-exploitation of, 19, 32, 34
communis estimatio, 52
compound specialization, 125–6, 128
conservative mentality, 5, 45, 130
consumption function, 108–10, 112
  stability of, 113
convertible husbandry, 80–1, 137–8
Crafts, N. F. R., 139
cultural continuity, 13, 20, 38, 59, 88, 131

Dahlman, C. J., 35, 45
de Vitry, Jacques, 51
Deane, P., and Cole, W. A., 137
defeudalization, 67, 72
  and diffusion of knowledge, 129
  and population density, 83, 88
  and technological progress, 60, 82, 88
demanorialization, 60, 82–3, 85, 129
demographic approach, 63, 71, 77–8, 81
Demsetz, H., 34
Derville, A., viii, 81
Desai, M., viii, 89
diffusion of knowledge, 9–10, 127
  and markets, 69, 72
  and servants, 69
  and size of the economy, 128
diminishing returns
  in agriculture, 69, 78, 84
  countervailing forces to, 71–3
  at low densities of population, 134
  and population pressure, 76, 118
distribution of income, 65, 72–3, 84, 109
division of labour, 10, 12, 56, 125–8, 130
  and aggregate demand, 40
  and indivisibilities, 11, 40
  in urban production, 76
  see also Smith, A.; specialization
Dobb, M., 63–4
Dodghson, R., 45
Domar, E., 83, 103

ecological equilibrium, 17
economies of practice, 9, 12, 124–5
economies of scale
  in grazing, 35, 47
  in urban production, 70
Elster, J., 5–6, 32
enclosures, 47
  and commercialization, 49
England
  agriculture, 80–1, 137
  population growth, 80
  traction power technology, 28–30, 85
exogenous shock–cultural diffusion paradigm, 13–15
externality, 6, 33
  definition of, 34
  in small societies, 35–6, 56
  see also partnership externality
Europe
  agriculture, independent evolution of, 20–1
  guilds, 50–4
  heavy soils, 27
  manorial production, 45
  regional diversity, 78–82
  stagnation, 134
  traction power, 29–30

Europe – *contd*
  windmills, independent evolution of, 28

Fenoaltea, S., 45, 47, 49
fertility, 87, 133
feudalism, 63–4, 67–8, 83
  and technological retardation, 84
feuds, 43–4, 60
Foreman-Peck, J., viii
formalist tradition, 42–3
France
  advanced agriculture, 76
  demanorialization, 82
  traction power, 30
  urbanization and self-sufficiency, 79
free-riders, 41
functional explanations
  in evolutionary biology, 58
  in historical materialism, 57
  of pre-industrial institutions, 59–60
Fussel, G., 25

Genoa, 54
German historical school, 42
gift exchange, 42–3
  and redistribution, 48
Goy, J., 104
Groth, C., viii
guilds
  and bargaining, 41
  as cartels, 50–1
  and fraud, 53
  and insurance, 54
  and transaction costs, 53
Gunnar from Hlidarend, 44

Habakkuk, H. J., 4, 63
Hahn, F., 61
Hall, A., 136

Hatcher paradox, 83, 89
Herlitz, L., viii
Hicks, J., 3
historical materialism, 5, 36–7, 57
hunter-gatherer societies, 3, 13–21, 34, 130–1

Icelandic sagas, 43–4
incentives for peasants and tenants, 65–6, 72, 84, 129
income
  foregone, 34
  net, 69, 84, 118
  per capita, 1–2, 65, 71–2, 75, 80, 107, 109–10, 132, 138–9
  and technological progress, 126–9
  see also aggregate demand; aggregate income
increasing returns, 70, 87
  and bargaining power, 40
independent evolution versus diffusion, 12–13
individualism, 36, 54–5
  and Calvinism, 55
  paradox of, 56
  and partnership externality, 56–7
  and property rights, 55
  restraint on, 40, 44
  and size of the economy, 57, 61, 70
indivisibility, 10–11, 31, 132
  in equipment and learning, 125
  in ploughing, 46
  see also division of labour
industrial revolution, 1, 4, 136, 140
institutional failure, 5–7, 19, 60, 130
institutions
  adequacy characteristics of, 37–8
  and chance variation, 58–9
  function versus origin of, 46
  and the nature of technology, 38–41
  and public goods, 60

institutions – *contd*
  and technological progress, 59–60, 130
insurance, 43, 47, 49, 54, 65, 77
investment in agriculture, 65, 69, 79, 82, 84
Italy, northern, agriculture, 27–8, 76, 78–9

joint product
  and monitoring of work effort, 40, 42, 49
  and partnership externality, 39, 48–9
Jörberg, L., viii
just price, 52

Kussmaul, A., 89

labour mobility, 68–9, 108
labour productivity, 18, 70
  in agriculture, 104, 111, 114–15, 128, 129
  decline in, 87
  in transition to agriculture, 18
Labrador Peninsula, 34
Labrousse, E., 104
land-holding classes, 65, 83, 108, 111, 116
land productivity, 105
land-saving technological change, 3, 12, 26, 39
  see also agriculture, rotation and fallowing regimes
Landes, D., 136
Langdon, J., 29–30
Le Goff, J., 52
Le Roy Ladurie, E., 4, 63, 71, 104
learning, by doing and using, 9
  see also economies of practice
Lefebvre de Noëtte, 29
London, 80

Low Countries
  agriculture, 27, 67, 76, 79, 85, 117, 137
  work effort, 81–2

McCloskey, D., 46–7, 49
Malthus, T., 63
Malthus–Ricardo interpretation of economic development, 88
  see also demographic approach
Malthusian equilibrium, 71–3, 134
  and technological progress, 86
Malthusian population growth, 65, 80, 94
manorial production, 45, 67, 83, 104–5
  and commitment of producers, 85
marginal productivity, 4, 19, 66
market, 2, 10, 42, 49, 68, 72, 83, 85
marriage pattern, north European, 86
Marshall, A., 9, 66
Marx, K., vii, 63
mediaeval village
  as a customary society, 3
  and individualism, 57
Mediterranean cultures, 131
Mercurey, vii
metallurgy
  in agriculture, 22, 24, 26, 77–8, 132
  Chinese, 21, 23
  development and chance variation, 21, 23, 30
  European, 21, 23
  independent evolution of, 21
  New World, 21
Mickwitz, G., 50–1
modernization
  and capitalist farming, 63–4
  and commercialization, 66
morals, 41, 44
mortality, 86–7
  and exogenous shocks, 133

Mumford, L., 136

Nedstam, B., viii
neolithic revolution, 13–21, 135
Njal, 44
Norfolk, 82
North, D., 6, 19, 34–5

open-field agriculture, 45, 59–60, 79

Palermo, 54
Parain, C., 24, 132
Pareto-optimality, 36
partible inheritance, 46
partnership externality, 39, 48, 56
patent rights, 6, 8
peasantry, 64, 83, 116
   and feudalism, 68
   innovations by, 85
   investment by, 69–70, 82
   legal status of, 83–4
   and specialization, 2
   taxation of, 84, 117
   see also Hatcher paradox
Persson, K. G., 139
Piggot, S., 14
Pirenne, H., 64
Polanyi, K., 42
population density, 11–12, 127
   and agricultural innovations, 81
   and defeudalization, 80
   and labour input, 81–2
population growth, 11–12, 135
   and defeudalization, 60, 66–7
   positive checks to, 86
   preventive checks to, 87
   and property relations, 65–7
   and substitution of labour for land, 16–17
population stagnation, 86–7, 133
population stress hypothesis, 17–18
Postan, M. M., 4, 63, 71, 77, 117

Postan–Le Roy Ladurie approach, 118
power, 44–5, 50, 56
   and equilibrium prices, 51
   as a source of externality, 35
pre-industrial societies, 1–3, 107
   institutions and technology in, 38–42, 60, 69, 124–7
   in Malthus–Ricardo trap, 4
private property
   and efficiency, 34
   versus collective rights, 35
private and social cost, 33, 34
property relations
   and distribution of income, 84
   and effort, 6, 55, 85
   and incentives, 65, 84
property relations approach, 63–4, 84, 99
property rights approach, 33–5
   and over-exploitation of the common, 19
   and transaction costs, 35–6
public goods, 41–2
purgatory, 52

Raaschou-Nielsen, A., viii
Radner, R., 39
random mutation and technological change, 7–8, 10, 69, 125–6
Reformation, 55
regional specialization, 12, 31, 66, 88
reinforcing mechanism, 58
Renfrew, C., 3, 21, 23
rent, 4, 69, 111, 114, 116
   and crop-sharing, 117
Rhine valley, agriculture in, 28, 79
Ricardian analysis, 65
Ricardo, D., 63
risk, 59
   and credit markets, 49
   insurance against, 46–7

risk – *contd*
  and size of the economy, 48
risk-sharing, 42
  and groups, 39
  and small economies, 40
  see also joint product; partnership externality
ritual, 43
Roman era, 56, 73, 131–3
Rome, 5, 131
Rosenberg, N., 4, 9, 136
Rotelli, C., viii

Sahlins, M., 19
Say's law, 92
Schoolmen, 51–2
Seebohm, F., 45–6
self-interested behaviour, 42
  and egalitarianism, 48
  and social continuity, 42
self-sustained growth, 7, 64, 67, 77, 118
separable tasks
  and division of labour, 125
  as indicators of technological progress, 131
serfdom, 66, 68
Sivéry, G., viii, 79
size of the economy, 48
  and individual freedom, 57, 70
  and level of technology, 127
  and political disintegration, 134
  and prices, 36
  see also individualism, paradox of
Skott, P., vii, 90
small economies
  and bargaining power, 39–40, 43
  and collusion, 53
  and restraints on individualism, 44–5, 56–7
Smith, A., 10, 50, 64, 125
Smith, R., vii, 38, 86

Smith effect, 70
social continuity, 38
  see also cultural continuity
specialization, 2, 66, 69, 73, 79, 91, 125, 127–8
  and enclosures, 47
  and increasing income, 71
  and urban growth, 64
stagnationist view, 1–3, 5, 33
Stone–Geary utility function, 91, 94
substantivist school, 42
Sweezy, P., 64

Tawney, R. H., 55
taxation, *see* rent
technical change
  and population pressure, 18–19
  as substitution, 2
  versus technological change, 1–2
technological change, 1–2, 3, 18, 37, 69, 71, 73
  demographic approach, 63, 78, 118
  endogenous, 8–13, 38
technological heritage, 10, 22, 127
  disintegration of, 130
  and inter-generational transmission of, 9, 38, 88, 124–5
echnological progress, 1, 4, 20, 38
  costless, 7, 9
  and diminishing returns, 78
  as growth of knowledge, 27, 126, 138–9
  intentional search for, 7
  and per-capita income, 126, 129–30
  and science 136, 140
technological sequences, 12–13
  complementarity of, 24–5
  fusion of, 22
tenantry, 64, 88
Thomas, R. P., 6, 34–5
Thrupp, S., 50–1

tithes, 104–6
Titow, J. Z., 77
trade, 2, 52, 64, 66, 69, 76, 138
transaction cost, 33, 51
trial and error, 10, 125–6, 140

urban proprietors, 67, 82
urbanization, 71, 74–6, 80–1, 108–10, 114
usury, 49, 53
utility theory, 52

van der Wee, H., viii, 117
van Houtte, J. A., 77
Vastrup, C., viii
Vinogradoff, P., 46, 48

watermills
  in China, 132
  and metallurgy, 133
  in Roman era, 5, 132
Weber, M., 55
welfare, 39, 58
  and exogenous constraints, 18
  and income, 2
  and productivity measurements, 113
Winchester estates, 77
woodwork, 125
Wrigley, E. A., viii, 119

yields, 81, 85, 87
  over an extended cycle, 106
  per seedcorn, 26, 105, 107
  per unit of land, 105